Uranium Resources on Federal Lands

Uranium Resources on Federal Lands

Charles F. Zimmermann
Foster Associates, Inc.

Lexington Books
D.C. Heath and Company
Lexington, Massachusetts
Toronto

Library of Congress Cataloging in Publication Data

Zimmermann, Charles F.
    Uranium resources on federal lands.

    Bibliography: p.
    Includes index.
    1.   Uranium industry—United States.
2.   Uranium mines and mining—United States.
3.   United States—Public lands.   I.   Title.
HD9539.U72U6   333.8′5   78-20738
ISBN 0-669-02847-9

    The study resulting in this publication was made under a fellowship granted by Resources for the Future. However, the conclusions, opinions, and other statements in this publication are those of the author and are not necessarily those of Resources for the Future.

    Additional support for this research was provided by Cornell University, under Grant No. AER74-21846 from the National Science Foundation. Any opinions, findings, and conclusions or recommendations expressed in this publication are those of the author and do not necessarily reflect the views of the National Science Foundation.

Published simultaneously in Canada.

Printed in the United States of America.

International Standard Book Number: 0-669-02847-9

Library of Congress Catalog Card Number: 78-20738

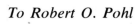
To Robert O. Pohl

Contents

List of Figures

List of Tables

Acknowledgments

This book concentrates on topics which will be of continuing interest in the future. Moreover, historical data have been collected for the purpose of illustrating long-term trends. It is hoped that the reader will find this study a useful guide to a broad range of literature on the topic of uranium supply, as well as a guide to specific issues regarding uranium leasing.

The author wishes to express his appreciation to Robert Kalter, Robert Pohl, Richard Schuler, and Robert Masson for their guidance and comments. Special thanks must go to John Broderick, who did the computer programming necessary to adopt the GEN2 model to uranium leasing; to Linda Skrabalak, who typed the entire manuscript; and to Joe Baldwin, who did all the drafting.

There are many individuals who helped to provide the information necessary for this research. Among those who have been particularly helpful are John Patterson, George Combs, J. Fred Facer, and William Voight, at the Department of Energy; John Martin and Philip Wood, at the Nuclear Regulatory Commission; Gerald Brubaker, at the Council on Environmental Quality; John Haring, at the Federal Trade Commission; Lewis Cook, at the Texas Department of Health Resources; Alphonso Topp, at the New Mexico Environmental Improvement Agency; Albert Hazle, at the Colorado Department of Health; John Hogerton, at the S.M. Stoller Corporation; Terry Lash, at the Natural Resources Defense Council; Peter Hayes, at Friends of the Earth Australia; Theresa Flaim, at Los Alamos National Laboratory; Stuart Harris, at the Australian National University; and Peter Auer, Donald Baker, Milton Leitenberg, and Edward Wonder, at Cornell University.

1 Introduction

In recent years there has been a great deal of controversy about the federal government's role in regulating the nuclear power industry. There is another aspect of federal involvement in the nuclear energy field which has received relatively little attention: the government's role as owner of uranium resources. Without uranium this nation's nuclear power production would grind to a halt. As of January 1974, over 75 percent of the uranium being produced in this country was mined from deposits discovered on lands owned by the federal government.[1] As of July 1, 1975, 59 percent of the acreage held for uranium exploration and mining was federally owned.[2] Thus the presence or absence of uranium exploration and mining activities on these and other federal lands with a potential for uranium discovery will have a major impact on the future of the nuclear power industry.

Corporations involved in uranium production are well aware of this situation. In 1973 the Nuclear Task Group of the National Petroleum Council expressed concern about the accessibility of federal land to uranium exploration:

> Approximately 50 percent of all proved and potential uranium resources are on federal or Indian lands; therefore, future land law changes and leasing policies could have a major impact on future uranium exploration activity.

> Federally controlled lands must be freely accessible for exploration if projected requirements for uranium from domestic reserves are to be met. All lands having uranium potential should remain available for exploration and development until exploration information allows assessment of mineral values.[3]

In 1974 the National Science Foundation sponsored a study of uranium resources by Battelle–Pacific Northwest Laboratories. Again, industry concern about access to federal lands was evident:

> Industry representatives contacted by Battelle generally expressed no desire for special government production incentives or subsidies. They were confident of their ability to meet uranium demand without preferential treatment from the government. However, they were particularly concerned with an increasing trend toward closure of federal lands to mining activity.[4]

1

In 1975 the National Academy of Science published a study of uranium resources. In this document there is an interesting comment by Samuel S. Adams, vice president of the Uranium Division of Anaconda Company:

> Perhaps the greatest problem related to the acquisition of mineral rights has to do with the public domain. It is certain that public lands will be subject to increased restrictions and substantially higher reclamation costs. It is also certain that tremendous time delays are likely to occur in the issuance of exploration rights, permits, exploration programs and other approvals now required by law.[5]

Taken together, these statements suggest that the government's role as owner of uranium resources will eventually receive more attention than it does today. The issue of uranium mining on government-owned lands also arises in Canada and Australia, both of which have a major share of the world's low-cost uranium reserves. Consequently, the way in which the United States deals with uranium development on federal land will set an example which will affect the course of policymaking in other countries. The topic is one of international significance.

The purposes of the present research are threefold:

1. To identify deficiencies in the present legal system regarding disposal of uranium on federal lands, and to outline alternative means for managing these lands.
2. To assess major factors which will influence uranium prices and production costs, and to develop projections of such values.
3. To analyze the relative merits of alternative systems for exploration, mining, and milling of uranium on federal lands.

The scope of this research involves the legal aspects of uranium development, the geological evidence underlying resource estimates, and the engineering aspects of uranium production and nuclear fuel requirements. These topics must all be covered in order to provide a sound basis for economic analysis. One of the key issues in uranium policy evaluation is the impact of uncertainty. In particular, the relative merits of alternative management systems depend strongly on the way in which uncertainty is faced by the public and private sectors.

Chapter 2 describes the federal government's role in uranium supply, in order to identify the various federal activities which influence uranium production. Chapter 3 discusses the legal situation in detail, with a view toward proposals for better management of the public's uranium resources.

Chapter 4 involves a discussion of domestic nuclear capacity projections and quantities of uranium required per unit of capacity. Chapter 5 provides a basis for estimating uranium mining and milling costs for the purpose of policy analysis. The impact of Nuclear Regulatory Commission licensing requirements on the cost of uranium production is evaluated. Chapter 6 is a discussion of various methods of resource estimation—principally, geological analogies and discovery-rate models—together with a description of present reserve characteristics, such as depth, deposit size, and grade. Chapter 7 reviews domestic price trends, methods of contracting for uranium supplies, and price-fixing allegations and antitrust considerations. Chapter 8 covers U.S. uranium imports and exports and world market trends.

Chapter 9 describes an analytical model which can be employed to compare the performance of different resource management systems under different price and cost scenarios. Uncertainty is incorporated into the model by means of Monte Carlo simulation techniques. Chapter 9 also includes a discussion of the impact of different leasing schedules on the domestic market. We have found, unfortunately, that publicly available data on uranium resources are presently inadequate to provide a reliable basis for calculating the impact of different schedules.

Notes

1. U.S. Federal Trade Commission Staff, *Report to the Federal Trade Commission on Federal Energy Land Policy: Efficiency, Revenue, and Competition* (Washington: USFTC, October 1975), p. 670A.

2. U.S. Energy Research and Development Administration, *Survey of Lands Held for Uranium Exploration, Development and Production in Fourteen Western States in the Six Month Period Ending June 30, 1975*, GJO-109 (76-1) (Grand Junction, Colo.: USERDA, December 1975), p. 2.

3. National Petroleum Council, *U.S. Energy Outlook: Nuclear Energy Availability* (Washington: National Petroleum Council, 1973), p. 7.

4. Battelle–Pacific Northwest Laboratories, *Assessment of Uranium and Thorium Resources in the United States and the Effect of Policy Alternatives*, PB-238 658 (Springfield, Va.: NTIS, 1974), p. 1.5.

5. National Academy of Sciences, Commission on Mineral Resources and the Environment (COMRATE), *Mineral Resources and the Environment, Supplementary Report: Reserves and Resources of Uranium in the United States* (Washington: National Academy of Sciences, 1975), p. 176.

2

Overview of Federal Activities Related to Uranium Exploration, Mining, and Milling

The focus of our research is the federal government's role as owner of uranium resources. Before this topic is covered, however, it is necessary to describe the sequence of steps by which uranium resources are converted into nuclear fuel.

The Nuclear Fuel Cycle

Steps in the Fuel Cycle

With one exception, all the commercial power reactors in the United States today are light-water reactors (LWRs). The one exception is a high-temperature gas reactor (HTGR) which is operated largely for research and development purposes.[1] Light-water reactors are supplied with enriched uranium, that is, uranium containing the isotope U-235 in a concentration which is higher than that found in nature.[2] The production of nuclear fuel for LWRs involves various steps, which are listed in table 2-1. The term used in industry and government to describe this sequence of steps is *the nuclear fuel cycle*. We will employ this term, since it is in common usage, but we feel that *nuclear fuel production process* would be more appropriate. The "cycle" is not really cyclical unless the uranium and plutonium contained in spent fuel are recycled. At present there are no commercial reprocessing plants operating in the United States, and no fuel is being recycled.

In order to evaluate federal policies regarding the disposal of uranium lands, we must examine at least the first step in the fuel cycle; that is, we must estimate the value of uranium ore reserves which have been or will be discovered on federal lands. The value of a reserve cannot be described simply by a dollar figure; one must examine different revenue streams associated with different production profiles. In other words, the mining process should be studied.

It is reasonable to ask which industries ought to fall within the scope of federal land policy, for any resource. Three considerations govern such a selection. First, in order to estimate the amount of economic rent the government can capture (as landowner) under different leasing systems, one must develop projections of costs and prices. Price projections will be

Table 2–1
The Nuclear Fuel Cycle for Light-Water Reactors

Step in the Nuclear Fuel Cycle	Product	Residuals	Ownership of Facilities Supplying Commercial Buyers	Production Sold Exclusively to Commercial Buyers
Exploration	Uranium ore reserves	(Minor)	Private	Yes
Mining	Uranium ore	Overburden and mine spoil	Private	Yes
Milling	Uranium concentrates or "yellow-cake," a solid containing U_3O_8	Tailings, which release radon, a gas	Private	Yes
Conversion to UF_6	Uranium hexafluoride, UF_6, a gas	(Negligible)	Private	No
Enrichment	Enriched UF_6 (high in U-235)	Tails (low in U-235)	Public	No
Conversion to UO_2	Brown oxide, UO_2, a solid	(Negligible)	Private	No
Production of UO_2 pellets	Pellets containing UO_2	(Minor)	Private	No
UO_2 fuel fabrication	Fuel rods containing pellets	(Minor)	Private	No
Reactor operation	Uranium and plutonium contained in spent fuel	Effluents and the remainder of the spent fuel	(Spent fuel is never sold)	No
Reprocessing	Uranyl nitrate, UO_2 $(NO_3)_2$, and plutonium nitrate, $Pu(NO_3)_4$, both liquids. These are subsequently converted to UF_6 and PuO_2, respectively.	High-level waste and other wastes	Private	No
Waste disposal	(None)	Various wastes	(Wastes are never sold)	

Note: The term *commercial buyers* refers to utilities and to companies which supply fuel to utilities. No commercial reprocessing plant is operating in the United States. One is under construction, but President Carter has said that it "will receive neither federal encouragement nor funding for its completion as a reprocessing facility" and has proposed to "defer indefinitely" the use of PuO_2 ("Carter Proposes To Ban the Use of Plutonium," *Wall Street Journal*, April 8, 1977, p. 2). Unless reprocessing is someday implemented, the uranium and plutonium contained in spent fuel will be a residual rather than a useful product.

very difficult to establish unless they refer to goods which are reasonably fungible. A price index for various grades and qualities of products is hard to develop when the relative quantities purchased change over time. Another consideration is that price projections are difficult to develop for products which are not marketed to a large degree, and for which a shadow price or imputed price must be estimated. The degree of vertical integration of an extractive industry, therefore, should affect the scope of resource leasing policy. Finally, an economically efficient level of rent capture cannot be determined unless upstream and downstream markets are workably competitive. Without competition, firms in these markets will be able to reduce the amount of economic rent accruing to landowners.

In the case of uranium resources, the first two considerations dictate that the milling industry must be studied as well as the mining industry. Although uranium concentrates are not all the same, they vary in quality far less than uranium ores. Moreover, concentrates are often sold by vertically integrated firms. Beyond milling, however, the necessary scope of land policy is less clear. There is a possibility, at least, that the upstream firms—the mining equipment suppliers and exploration drilling contractors—will set artificially high prices for their goods and services, whereas the downstream firms—the conversion and nuclear fuel fabrication industries—may set artificially low prices for uranium concentrate.

Clearly there is a need for leasing policy to be coordinated with antitrust policy. However, we have chosen not to investigate industries other than mining and milling, primarily because a study of these industries would require too great a research effort, and because there is no conflict between the long-term objectives of antitrust and leasing policies, even though attempts to promote competition via divestiture will cause short-term disruptions.

Definitions of Uranium Ore and Yellowcake

No clear definition of *ore* exists in the uranium mining industry. *Uranium ore* is not among the twenty-nine legal terms defined by the Atomic Energy Act of 1954.[3] Industry usage of the term *ore* conforms to the dictionary definition: "any natural combination of minerals, especially one from which a metal or metals can profitably be extracted."[4] The profitability of extraction obviously varies with price and cost variations.

Any attempt to define *ore* must be abandoned entirely in the case of recovery of uranium *in situ*:

> Where traditional methods are used there are actually two points in the process from extraction through concentration at which there exists a

marketable product; first where crude ore is severed from the ground, and later where concentrate is produced. In a solution mining operation this is not true.[5]

Interest in solution mining has increased recently. However, it raises a number of legal problems related to land subsidence, extraction of uranium beyond the boundary of a property, pollution of ground water, and operation at low recovery rates.[6] As of October 1976, its use was not widespread: "Production from solution mining has been relatively constant at less than 1 percent of the total uranium production for more than 15 years. This may increase slightly in 1976."[7]

Neither ore nor yellowcake is a uniform, fungible product. A high-grade ore can contain as much uranium as a low-grade yellowcake, so the two commodities form a sort of continuum in terms of uranium concentration. The uranium ore which was imported from the Belgian Congo to supply the Manhattan Project contained 65 percent U_3O_8.[8] In 1970 an Australian mining company reported discovery of a "massive" pitchblende lode averaging 65 percent U_3O_8.[9] Yellowcake in the U.S. market may contain as little as 70 percent U_3O_8, although average concentrations are higher:

> U_3O_8, an uranium-uranyl oxide, is olive green in color and contains 84.8 percent U. . . .
>
> "Yellowcake," the term applied to the concentrate produced at mills, is generally considered to be ammonium diuranate $(NH_4)_2U_2O_7$ or sodium diuranate $(Na_2U_2O_7)$. The exact composition is variable and depends upon precipitating conditions. Refinery specifications for U_3O_8 grade, elemental composition, and impurities in yellowcake have been established by ERDA [Energy Research and Development Administration] and industry. ERDA specifications require a minimum of 75 percent U_3O_8; minimum industry specifications range from 70 to 77 percent U_3O_8. However, most concentrates contain a minimum of 75 percent U_3O_8 and average 80 to 85 percent U_3O_8. In addition, the concentrate must be amenable to refining by ion exchange or solvent extraction.[10]

Thus yellowcake averages about 70 percent uranium, and may contain as little as 59.4 percent uranium. Concentrate prices are measured in terms of dollars per pound of U_3O_8, and quantities are measured in terms of their U_3O_8 content. Uranium reserves are measured in thousands of tons of U_3O_8 contained in ores which can be mined and milled at or below a certain cost level.[11]

When yellowcake is converted into uranium hexafluoride, it becomes a fungible product. All uranium hexafluoride must be delivered to federally owned enrichment plants and must meet strict standards of purity.[12]

The U-235 content of domestic ores varies slightly, from 0.7103 to

0.7113 weight percent, but Department of Energy (DOE) enrichment plants accept UF_6 from these ores on a uniform basis, without regard to these variations. The Atomic Energy Commission (AEC) established an accepted value of 0.711 percent, effective July 1, 1963. However, ores from Gabon have been found to contain as little as 0.44 atom percent U-235, due to chain reactions which occurred 2 billion years ago. If a domestic ore deposit were found to be similarly depleted, it would not yield fungible UF_6.[13]

Economies of scale in conversion facilities dictate that very few milling companies can be vertically integrated forward. At present there is only one milling company—Kerr-McGee—which operates a conversion facility.[14]

Management of Uranium Lands

Alternative Methods of Managing Uranium Resources

The federal government owns a wide variety of different lands, which are usually classified according to the type of ownership (method of acquisition, surface rights, trust responsibilities, etc.) and the federal agency holding primary responsibility for use of the land. The fundamental role of the legal system governing the management of energy resources is to specify, for a given category of land, the method by which these resources may be extracted. To date, there are four alternative methods which have been applied to uranium resources.

1. Private exploration and development (claims system). The government has enacted legislation which allows private individuals and corporations to claim minerals on a "first come, first serve" basis. The first to discover and claim a valuable mineral deposit becomes the first to own it.

2. Private exploration and development (leasing system). The government has sold rights to explore public lands and to extract and process minerals where exploration has led to the discovery of valuable deposits. In some cases the right to explore public land has been issued without charge via a mineral permit. Private corporations own and operate all the facilities involved.

3. Government exploration and development. The government has owned facilities for exploration, mining, and processing of energy resources, but it has awarded contracts to private corporations for operation of these facilities.

4. Government exploration, private development. The government has conducted exploration programs, with or without awarding contracts

to private corporations, and then sold rights to extract and process minerals from discovered deposits.

The third option is no longer being employed. The second and fourth options can each be further subdivided into three categories: situations in which two or more bidders offer competitive bids to purchase rights, situations in which only one bidder offers a bid although other bids are permitted, and situations in which the sale of rights is negotiated with only one firm. The term *competitive sale* is sometimes used to describe situations with only one bidder, as well as those with two or more bidders. We find this terminology to be misleading.[15] Since the federal government is heavily involved in uranium exploration, both directly and indirectly, the distinction between private exploration and development and government exploration with private development depends on the precise way in which one chooses to define these terms. In order to discuss the various types of federal land which can be leased, it is most convenient to lump these two management alternatives together under the general category of uranium leasing.

Minerals encompassed by a mining claim can be purchased from the federal government by means of a patient. A patented claim conveys legal title to minerals (and to surface resources, in some instances), while an unpatented claim conveys only the right to extract minerals during the time period in which the claim is kept active by ''annual assessment work''—the expenditure of at least $100 per claim year on labor and improvements.[16] It is possible for the government to sell mineral right to one individual via a mineral patent and surface rights to another via a land patent on a single tract of land. The government receives no revenue from unpatented claims.

Agencies Involved in Management of Uranium Lands

The Bureau of Land Management (BLM), within the Department of the Interior, administers all mineral-development activity on lands subject to uranium mining claims. In addition, BLM administers the surface resources on the major proportion of these lands. To date, the minimg claim system (or ''location-patent system'') has been the dominant method of disposing of uranium resources. It is quite probable that BLM will continue to manage resources on what are now claim lands even if the claim system is replaced by a leasing system. In addition, BLM is already involved in leasing uranium resources in portions of the national forest system.

The Bureau of Indian Affairs administers the disposal of uranium resources on Indian lands, which have made a large contribution to domestic uranium production. In a sense, Indian lands are not federal lands, because they are not administered for the benefit of the general public. However, they fall within the scope of the Bureau of Land Management's definition of federal land, namely, "all classes of land owned by the federal government:"[17]

> Legal title to Indian lands is vested in the United States Government. The doctrine of federal ownership originated not in the Constitution or any federal statute or treaty, but in the 1823 Supreme Court decision of *Johnson and Graham's Lessee v. McIntosh*, 21 U.S. (8 Wheat.) 240 (1823).[18]

Leasing is the method of disposal of uranium resources on all Indian lands except the Papago Indian Reservation, which is open to mining claims.

The Forest Service, within the Department of Agriculture, administers the surface resources of national forests. Most of the national forest system is open to mining claims, but a portion is subject to mineral leasing. In either case, uranium exploration, mining, and milling must be regulated to avoid damage to forests and wildlife.

The Department of Energy (DOE) administers certain uranium land leases which were issued by the Atomic Energy Commission. Although DOE possesses broad powers to lease uranium-bearing lands, it is unlikely that the agency will initiate a leasing program in the near future.

The National Park Service administers certain lands which are open to uranium mining claims. However, national parks are generally closed to all types of mining activity.

The Bureau of Reclamation reviews applications to have specific reclamation lands opened to mining claims. If mining activity on these lands will not interfere with necessary irrigation projects, then claims will be permitted. The Bureau of Land Management is also involved in this review process. No substantial contribution to domestic uranium production is involved.

Our discussion here is intended to provide only a general overview of disposal activities. We will return to the topic, providing greater detail, in chapter 3.

Participation in Uranium Exploration

Direct Federal Involvement

One form of government activity which deserves close scrutiny is direct participation in uranium exploration. In the past 5 years, substantial changes have occurred in the way AEC, ERDA, and DOE officials have viewed their responsibilities in this area. In January 1973, an AEC official described the exploration programs conducted by private companies in terms of three phases. The first phase consisted of airborne radiometric surveys and geologic studies covering an area "up to tens of thousands of square miles;" the second phase involved land acquisition and wide-spread drilling over an area "up to a few hundred square miles;" and the third phase consisted of close-spaced drilling and detailed exploration of an area "of a few square miles."[19]

In March 1973, the AEC's assistant director for raw materials announced a new federal program—an evaluation of the "long-range potential" for domestic uranium supplies. Emphasis was placed on areas which were not currently being explored by private sector. The problem was analogous to a research and development project which would have long-term benefits; the government would ask the industry for "cooperation":

> We in AEC intend largely to emphasize an assessment of the total resource potential of the country, including an attempt to evaluate the possibilities for finding uranium in areas not previously investigated or in which economic amounts of uranium have not yet been detected. We hope to develop a preliminary assessment of the total uranium potential in the U.S. in the next two or three years, and to have a somewhat more quantitative evaluation at the end of 1978 [sic].

> Our objective is a continuing study which, after the initial report, will result in an annual updating, rather than an intense, one-time operation. We look forward to the further cooperation of the industry, and believe that in time this effort should provide a useful and comprehensive evaluation of the United States' prospects for a long-term uranium supply.[20]

Three years later, ERDA described the program to the Joint Committee on Atomic Energy in somewhat different terms: "The purpose of the program is to develop new information to stimulate and assist industry in exploring for uranium in the United States."[21] Moreover, ERDA officials stated that the agency should, in effect, take responsibility for ensuring that the total funding of uranium exploration in the country is sufficient to prevent reserve shortages: "We believe that strong early government programs to identify new favorable exploration areas will stimulate larger industry expenditures, will be the most cost effective approach and will avoid the necessity for heavier government spending later."[22]

Comparison with Industry Expenditures

The annual cost of the program has grown rapidly, as Table 2–2 shows. Geologic research, airborne radiometric surveys, and drilling projects are

Table 2-2
Annual Funding of Uranium Exploration (Other than Land Acquisition and Drilling Done by Private Companies) and Uranium Ore Processing Research Conducted by the Bureau of Mines

Calendar Year	Industry Exploration Expenditures, Other than Land Acquisition and Drilling ($ million)	Fiscal Year	AEC/ERDA "Source Materials" and "Uranium Resource Assessment" Budget ($ million)	Geological Survey "Uranium/Thorium Resource Assessment" Budget ($ million)	Bureau of Mines "Uranium Ore Processing" Budget ($ million)
1972	9.60[a]	1973	1.924[c]		
1973	16.53[a]	1974	2.770[d]	1.1194[f]	0.199[g]
1974	21.71[a]	1975	6.730[e]	4.263[f]	1.105[g]
1975	31.52[a]	1976	14.8 (est.)[e]	4.302 (req.)[f]	1.115 (req.)[g]
1976	47.79[b]	1977	30.0 (est.)[e]	4.915 (req.)[h]	1.127 (req.)[i]

Note: Agency estimates of budget outlays are marked "est." and agency requests for appropriations are marked "req."

[a]William L. Chenoweth, "Exploration Activities," *Uranium Industry Seminar*, GJO-108(76) (Grand Junction, Colo.: USERDA, 1976). p. 180.

[b]USERDA, *Uranium Exploration Expenditures in 1976 and Plans for 1977–78*, GJO-103(77) (Grand Junction, Colo.: USERDA, 1977), p. 1.

[c]U.S. Congress, Joint Committee on Atomic Energy, *AEC Authorizing Legislation, Fiscal Year 1975*, Hearings, 93d Cong., 2d Sess., February 19, 26, 27, 28, March 4 and 5, 1974 (Washington: GPO, 1974), pp. 331, 348.

[d]U.S. Congress, Joint Committee on Atomic Energy, *ERDA Authorizing Legislation, Fiscal Year 1976*, Hearings, 94th Cong., 1st Sess., February 4, 18, 27, March 4, 6, 11 and 13, 1975 (Washington: GPO, 1975), pp. 673, 716.

[e]U.S. Congress, Joint Committee on Atomic Energy, *ERDA Authorizing Legislation, Fiscal Year 1977*, Hearings, 94th Cong., 2d Sess., January 21, February 4, 17, March 11, 17, 18, and 19, 1976 (Washington: GPO, 1976), p. 710. These figures represent budget outlays. FY 1976 ends June 20, 1976 and FY 1977 begins August 1, 1976. During the "transition quarter," the uranium resource assessment program involved budget outlays estimated to be $4.824 million. This sum has not been added to the figures in the table, which refer to 12-month periods for purposes of comparison.

[f]U.S. Congress, House, Committee on Appropriations, Subcommittee on Department of the Interior and Related Agencies, *Department of the Interior and Related Agencies Appropriations for 1976*, Hearings, 94th Cong., 1st Sess. (Washington: GPO, 1975), Part 5, p. 650.

[g]Ibid., Part 6, p. 99.

[h]U.S. Congress, House, Committee on Appropriations, Subcommittee on Department of the Interior and Related Agencies, *Department of the Interior and Related Agencies Appropriations for 1977*, Hearings, 94th Cong., 2d Sess. (Washington: GPO, 1976), Part 1, p. 504.

[i]Ibid., Part 1, p. 708.

conducted.[23] In effect, the federal government is shouldering an increasing proportion of the cost of both regional and detailed reconnaissance.[24] Reports issued by ERDA in 1976 include only one summary of a drilling project, so drilling apparently consumed only a small proportion of ERDA and DOE expenditures. Thus federal funding for uranium resource assessment should be compared with industry expenditures, exclusive of land acquisition and drilling. As Table 2–2 shows, federal expenses for fiscal year 1977 exceed industry expenses for calendar year 1975.

Even if one compares these federal expenses with the total value of 1977 domestic delivery commitments, the level of federal funding is significant. As of July 1, 1976 there were 13,800 tons U_3O_8 committed for delivery in 1977 by domestic producers to domestic buyers. Ninety-one percent of this tonnage was under contract prices, and prices for 80 percent of the contract-price tonnage were reported to ERDA. If we apply the average reported price to total delivery commitments, we get a value of $348 million. As Table 2–2 shows, federal funding for uranium and thorium resource assessment in FY 1977 is estimated to be $34.9 million, or about 10 percent of the value of 1977 deliveries of U_3O_8. Since there are only two power reactors in the United States (operable, under construction, or on order) which run on thorium (that is, Fort St. Vrain and Shippingport), the United States Geological Survey (USGS) budget for thorium assessment is likely to be very small in relation to the budget for uranium.[25]

Indirect Federal Involvement

In addition to direct participation, the federal government subsidizes uranium exploration indirectly, in four ways:

1. A depletion allowance equal to 22 percent of gross income from mining and milling or 50 percent of taxable income, whichever is smaller, is applied to calculate income subject to federal taxes.[26]
2. Development expenditures can be deducted from current income in order to compute taxable income for federal tax purposes. Development drilling costs qualify for this deduction.[27]
3. Exploration expenditures can be deducted from current income for federal tax purposes if these expenditures are either added to gross income in the year when production begins or else deducted in such a way that they offset the depletion deduction.[28]
4. Firms can be reimbursed for exploration expenditures by the Office of Mineral Exploration of the Department of the Interior for up to 50 percent of authorized costs.[29]

Note that the depletion allowance subsidizes all phases of uranium mining and milling, even though its primary rationale is that it makes funds available for exploration to replace depleted reserves with new reserves. Also note that development drilling is considered an exploration expenditure by DOE[30] and is part of the third phase of exploration among the three phases described by the AEC, yet it is not considered an exploration expenditure for federal tax purposes. In tax terminology, development expenditures are "all expenditures paid or incurred during the taxable year for the development of a mine or other natural deposit (other than oil or gas well) if paid or incurred after the existence of ores or minerals in commercially marketable quantities has been disclosed."[31]

Need to Coordinate Policies Regarding Exploration and Disposal of Uranium Lands

Where federal lands are involved, the effect of all these direct and indirect programs to support uranium exploration is to bring resource management procedures closer to one of the alternatives we discussed on page 9, namely, federally sponsored exploration followed by leasing of the most promising areas. However, leasing is not presently conducted with this alternative in mind. Among existing leases, only those signed by the AEC in 1974 on lands explored during the 1950s were intentionally based on this type of resource-management policy.[32] No analysis has been made of the impact of recent federal assistance in exploration on present or prospective leasing programs. One reason for this lack of analysis is the division of responsibilities among federal agencies. DOE, USGS, the Internal Revenue Service (IRS), and the Office of Mineral Exploration implement policies which fund or assist exploration; BLM, the Forest Service, and the Bureau of Indian Affairs are the agencies which play the most important roles in administering uranium-bearing federal lands. The Forest Service is involved by virtue of the fact that mining claims can be made in national forests on the public domain, and mineral leases are sometimes awarded in national forests on acquired lands.

Need to Coordinate Domestic and Foreign Policy Regarding Uranium Supplies

In the case of oil production on federal lands—particularly on the outer continental shelf—there is a well-known tradeoff between environmental objectives and the need to reduce our reliance on imports. With regard to uranium production on federal lands, the same conflict of goals exists, but it is a relatively long-term issue. Clearly, utilities will have to increase their reliance on uranium imports if mining companies are denied access to domestic lands containing uranium, or if the construction of mines and mills is prevented or delayed in order to protect the environment or the public health and safety. However, our present reliance on imports is modest. As table 2–3 shows, imports represented 10.8 percent of past deliveries and contract commitments at the end of 1977.

It would not be wise to conclude, on the basis of table 2–3, that the United States will be practically self-sufficient in uranium in the future. Only 107,700 short tons of U_3O_8 had been delivered by domestic producers to domestic buyers as of January 1, 1978.[33] The bottom line of table 2–3 includes an additional 193,000 tons committed for future delivery and subject to controversies over environmental and safety hazards. Many hundreds of thousands of tons of additional supplies will be needed to fill the

Table 2–3
Data on the Total Quantity of Uranium Involved in Past Deliveries and Contract Commitments for Future Deliveries to Commercial Buyers in the United States.
(in thousand tons U_3O_8)

	Domestic Producers			Foreign Producers		
Year	New Purchases	Reductions	Net Change	New Purchases	Reductions	Net Change
1/1/66–2/1/71			108.6			
1971[a]	10.0	4.5	5.5			
1972	19.7	4.0	15.7			8.0
1973	52.0	6.2	45.8			
1974	17.6	1.9	15.7			33.0
1975	16.2	1.7	14.5	4.8	0.4	4.4
1976	92.9	9.5	83.4	1.8	0	1.8
1977	12.0	.5	11.5	1.5	12.3	−10.8
Totals			300.7			36.4

Sources: U.S. Atomic Energy Commission, *Nuclear Industry Fuel Supply Survey,* WASH-1196 (Washington: USAEC, 1972), p. 2; USAEC, *Survey of United States Uranium Marketing Activity,* WASH-1196(73) (Washington: USAEC, 1973), p. 2; USAEC, *Survey of United States Uranium Marketing Activity,* WASH-1196(74) (Washington: USAEC, 1974), p. 2; U.S. Energy Research and Development Administration, *Survey of United States Uranium Marketing Activity,* ERDA-24 (Washington: USERDA, 1975), pp. 2, 7; John A. Patterson, "Uranium Market Activities," *Uranium Industry Seminar,* GJO-108 (76) (Grand Junction, Colo.: USERDA, 1976), p. 29; John A. Patterson, "Uranium Market Activities," *Uranium Industry Seminar,* GJO-108(77) (Grand Junction, Colo.: USDOE, 1977), p. 38; U.S. Department of Energy, *Survey of United States Uranium Marketing Activity,* DOE/RA-0006 (Washington: USDOE, 1978), pp. 2,15–16.

[a]Data for the month of January 1971 are not included here.

30-year requirements of domestic reactors in operation, under construction, and on order. As of September 14, 1978, the amount of capacity committed was 199,068 MWe (megawatts, electrical); assuming 7 short tons (ST) of U_3O_8 per MWe, lifetime requirements would be about 1.4 million tons.[34] If uranium supplies become available from developing countries at prices substantially below domestic prices, safety concerns will tend to increase our reliance on imports. Some means must be devised to coordinate domestic and foreign policy on this issue.

If the United States became seriously concerned about the economic incentives to recycle plutonium in foreign reactors, it would be necessary to consider becoming a net exporter of uranium. The critics of nuclear power face a choice between the lesser of two evils—uranium mining or plutonium recycling. There is a potential for intense conflict between domestic and foreign policy objectives in this context.

Notes

1. Lists of reactors subject to construction permit and licensing proceedings are published regularly in *Nuclear Safety*, a bimonthly journal of the U.S. Nuclear Regulatory Commission.

2. A good introduction to the principles of reactor operation can be found in David R. Inglis, *Nuclear Energy: Its Physics and Its Social Challenge* (Reading, Mass.: Addison-Wesley, 1973).

3. 42 U.S.C. 2014.

4. *Webster's New World Dictionary* (New York: William Collins & World Publishing Co., 1974), p. 1001.

5. Royal E. Peterson, "The Uranium Royalty Provision: Its Evolution, Present Complexity and Future Uncertainty," *Rocky Mountain Mineral Law Institute* 22 (1976):865–889.

6. Thomas E. Root, "Legal Aspects of Mining by the In Situ Leaching Method," *Rocky Mountain Mineral Law Institute* 22(1976):354–362.

7. J. Fred Facer, Jr., "Production Statistics," *Uranium Industry Seminar*, GJO-108(76) (Grand Junction, Colo.: USERDA, 1976), p. 194.

8. Richard G. Hewlett and Oscar E. Anderson, Jr., *The New World* (Washington: U.S. Atomic Energy Commission, 1972), p. 26.

9. "Australia Claims Largest, Richest Uranium Deposit Yet Discovered," *Nuclear Industry*, September 1970, pp. 33–34.

10. Walter C. Woodmansee, "Uranium," in *Mineral Facts and Problems*, U.S. Department of the Interior, Bureau of Mines (Washington: GPO, 1976), pp. 1180–1181.

11. USAEC, *Nuclear Fuel Resource Evaluation: Concepts, Uses, Limitations*, GJO-105 (Grand Junction, Colo.: USAEC, 1973), pp. 7–9;

OECD Nuclear Energy Agency and International Atomic Energy Agency, *Uranium: Resources, Production and Demand* (Paris: OECD, 1976), pp. 11–12.

12. Specifications for UF_6 delivered to DOE for commercial use were originally published in the *Federal Register* on November 29, 1967 (at 32 FR 16289). This notice has been revised many times by subsequent notices. *The Nuclear Regulatory Reporter*, paragraph 1000, contains the most recent version of the UF_6 criteria.

13. Walter C. Woodmansee, "Uranium," in *Mineral Facts and Problems*, Department of the Interior, Bureau of Mines (Washington: GPO, 1976), p. 1180; George A. Cowan, "A Natural Fission Reactor," *Scientific American*, July 1976, pp. 36–47.

14. Arthur D. Little, Inc., *Competition in the Nuclear Power Supply Industry* (Cambridge, Mass.: Arthur D. Little, 1968) pp. 176–182. A list of companies operating fuel cycle facilities other than mines and mills is published annually in the *Minerals Yearbook*, issued by the Department of the Interior, Bureau of Mines.

15. U.S. Federal Trade Commission Staff, *Staff Report on Mineral Leasing on Indian Lands* (Washington: GPO, 1976), pp. 26–27, 60.

16. U.S. Federal Trade Commission Staff, *Report to the Federal Trade Commission on Federal Energy Land Policy: Efficiency, Revenue, and Competition* (Washington: GPO, 1976), pp. 658–661. See also 43 CFR 3850.

17. U.S. Department of the Interior, Bureau of Land Management, *Public Land Statistics, 1975* (Washington: GPO, 1976), p. 36.

18. U.S. Federal Trade Commission Staff, *Staff Report on Mineral Leasing on Indian Lands* (Washington: GPO, 1976), p. 3.

19. Elton A. Youngberg, "The Uranium Industry: Exploration, Mining, and Milling," *IEEE Transactions on Power Apparatus and Systems* (July-August 1973): 1203.

20. Robert D. Nininger, "Uranium Reserves and Requirements," *Nuclear Fuel Resources and Requirements*, WASH-1243 (Washington: USAEC, 1973), pp. 23, 25.

21. U.S. Congress, Joint Committee on Atomic Energy, *ERDA Authorizing Legislation*, Fiscal Year 1977, Hearings, 94th Cong., 2d Sess., January 21, February 4 and 17, and March 11, 17, 18, and 19, 1976 (Washington: GPO, 1976), p. 2832.

22. Ibid., pp. 2826–2827.

23. Robert C. Malan, "Status of the NURE Program," *Uranium Industry Seminar*, GJO-108(76) (Grand Junction, Colo.: USERDA, 1976), p. 67.

24. For example, one ERDA report is titled, "The Stratigraphy and Environments of Deposition of the Cretaceous Hell Creek Formation

(Reconnaissance) and the Paleocene Ludlow Formation (Detailed), Southwestern North Dakota," GJBX-22(76) (Grand Junction, Colo.: USERDA, 1976).

25. See John A. Patterson, "Uranium Market Activities," *Uranium Industry Seminar*, GJO-108(76) (Grand Junction, Colo.: USERDA, 1976), pp. 32, 43, 48, 49; Roman V. Sondermayer, "Thorium," *Minerals Yearbook, 1973*, U.S. Department of the Interior, Bureau of Mines (Washington: GPO, 1976).

26. Clayton J. Parr and Northcutt Ely, "Mining Law," in *SME Mining Engineering Handbook*, edited by Arthur B. Cummins and Ivan A. Given (New York: Society of Mining Engineers, 1973), pp. 2-33 to 2-37; National Petroleum Council, *U.S. Energy Outlook: Nuclear Energy Availability* (Washington: National Petroleum Council, 1973), pp. 58–59, 107–108.

27. Ibid.

28. Ibid.

29. Walter C. Woodmansee, "Uranium," in *Mineral Facts and Problems*, U.S. Department of the Interior (Washington: GPO, 1976), p. 1194. See also 30 C.F.R. 229, the USGS regulations concerning federal assistance in financing exploration for mineral reserves, excluding organic fuels.

30. For example, see U.S. Energy Research and Development Administration, *Exploration Expenditures in 1976 and Plans for 1977–78*, GJO-103(77) (Grand Junction, Colo.: USERDA, 1977).

31. Parr and Ely, "Mining Law," p. 2-37.

32. "First Leases on AEC Uravan Lands Up for Bidding," *Nuclear Industry*, August 1973, pp. 34–35; "Uravan Pre-Bid Meeting," *Nuclear Industry*, January 1974, p. 40.

33. U.S. Department of Energy, *Survey of United States Uranium Marketing Activity*, DOE/RA-0006 (Washington: USDOE, 1978), p. 3.

34. Atomic Industry Forum, *Nuclear INFO*, September 1978, p. 4.

3

Legal Aspects of Uranium Development on Federal Lands

We have seen that the federal government influences uranium supply in many different ways. In this chapter we shall examine two kinds of activity in greater detail: leasing and environmental regulation. These activities are very closely related, and it is impossible to give an accurate picture of leasing policy without also discussing environmental policy. Moreover, in order to understand the effects of leasing, one must examine the alternative which prevails today—the mining claim system.

The discussion begins with a description of the present legal system for managing uranium resources on federal lands. Next, an introduction to federal environmental regulation of uranium mining and milling is provided. Since a portion of federal authority has been transferred to certain states, and since uranium development is conditioned upon regulatory approvals from state governments, a review of state environmental regulation is presented. Finally, proposals for changing the present legal system at the federal level are discussed.

The Present Legal System

Types of Federal Land

All land owned by the federal government in the fifty states can be divided into two categories: public-domain land and acquired land. *Public-domain lands* include "lands which never left Federal ownership; also, lands in Federal ownership which were obtained by the Government in exchange for public [domain] lands or for timber on public [domain] lands."[1] *Acquired lands* have "been obtained by the Government through purchase, condemnation or gift, or by exchange for purchased, condemned or donated lands, or for timber on such lands."[2] The distribution of acquired land by federal agency is shown in table 3–1. The aggregate of public-domain land, acquired land, the Canal Zone, and every other kind of land owned throughout the world by the federal government is called *federal land*, whereas the term *public land* is simply a short form of *public-domain land.*[3]

There is one category of federal land, within the fifty states, which must be considered separately in the context of public policy: Indian

Table 3–1
Distribution of Acquired Land, by Federal Agency, as of June 30, 1974

Agency	Area (acres)	Percentage of Total
Department of Agriculture		47.9%
Forest Service	27,053,951.5	
Other	91,419.4	
Department of the Interior		24.5%
National Park Service	4,984,458.6	
Fish and Wildlife Service	3,920,428.6	
Bureau of Land Management	2,351,890.1	
Bureau of Reclamation	1,868,169.1	
Bureau of Indian Affairs	764,201.4	
Other	27,153.8	
Department of Defense		24.1%
Corps of Engineers	7,070,249.9	
Army	3,943,788.0	
Air Force	1,424,859.0	
Navy	1,250,586.1	
Tennessee Valley Authority	924,660.2	1.6%
Atomic Energy Commission[a]	667,117.7	1.2%
Other agencies	376,184.7	0.7%
Total	56,719,118.1	100.0%

Source: U.S. Department of the Interior, *Public Land Statistics, 1975* (Washington: GPO, 1976), pp. 14–31.

[a]On January 19, 1975, these lands came under the control of the Energy Research and Development Administration. On October 1, 1977, they came under the control of the Department of Energy.

reservations. Either public-domain land or acquired land can be reserved for Indians. The reason why this land should be considered separately is that the criterion governing its use is not the best interests of the federal government or the people of the United States as a whole, but the best interests of the Indians. In theory, federal energy policy should not apply to Indian land, except at the convenience of the Indians.[4] The American Indian Policy Review Commission has issued a "tentative final report" recommending that the federal government exert even less control over Indian land than it does today, and that the Bureau of Indian Affairs be abolished.[5] Moreover, from an ethical standpoint, one can question whether these lands should even belong to the federal government—it can be argued that they should belong to the Indians. In view of all these controversial issues, it is best to regard policy toward uranium resources on Indian reservations as distinctly different from policy toward uranium resources on other federal lands.

A significant portion of federal land (other than Indian land) is already being used for purposes which are inconsistent with uranium exploration and development. In addition, there will always be some portion which does not offer enough geological evidence of uranium availability to warrant exploration drilling. Of the remainder, public-domain land is subject to either leasing or claims, and acquired land is subject only to leasing.[6] The various categories of land involved are shown in table 3–2. The following discussion will simply fill in the details of the scheme outlined in this table.

Lands Subject to Uranium Leasing

Indian Reservations. There are three categories of Indian land on which federally owned mineral resources can be leased. Two sources of statutory authority are involved. The Bureau of Indian Affairs is responsible for such leasing.

1. Allotted land is held by the U.S. government in trust for the benefit of an individual Indian. It is leased under the act of March 3, 1909, 25 U.S.C., Sec. 396. The regulations which implement this statute are found in 25 C.F.R., Part 172.
2. Tribal land is held by the United States in trust for an entire tribe. It is leased under the Omnibus Mineral Leasing Act of May 11, 1938, 25 U.S.C. Section 396a–f, which is implemented through 25 C.F.R. Part 171.
3. Ceded land was ceded to the federal government and settled by non-Indians while the tribe retained the mineral rights in its name. It is leased under the same statute and the same regulations as tribal land.[7]

Lands Administered for National Park, Monument, and Wildlife Purposes. These can be leased by DOE "only when the President by Executive Order declares that the requirements of the common defense and security make such action necessary."[8] This authority is granted by the Atomic Energy Act of 1954, but no regulations have been published to implement it because the AEC, ERDA, and DOE have never recommended that the President make such a declaration.

Other Leasable Public-Domain Land. All public-domain land is open to mining claims unless declared otherwise, that is, unless "withdrawn from entry and location under the general mining laws."[9] An *entry* is "an application to acquire title to public lands," for example, an application for a patented mining claim. A *location* consists of all the things one must

Table 3-2
Different Categories of Federal Land, Grouped According to the Type of Leasing or Claims System Applied to Uranium Resources

	Public Domain	*Acquired Land*
Subject to Leasing	Indian reservations	
	Lands administered for national park, monument, and wild-life purposes[a]	
	Lands withdrawn for use by DOE	National forest lands
	Lands withdrawn for other uses, but leased by DOE[b]	Other lands leased by the Bureau of Land Management
		Lands leased by DOE but not controlled by DOE[b]
		Lands leased and controlled by DOE
Subject to Claims	Vacant public-domain lands	
	Lands containing Leasing Act minerals	
	National forest lands	
	Lands which have been sold or conveyed with the mineral rights reserved to the United States	
	Lands which have been withdrawn, without precluding claims	
	Lands containing uranium-bearing lignite	
	Certain other lands[c]	

[a]These can be leased by DOE "only when the President by Executive Order declares that the requirements of the common defense and security make such action necessary" (42 U.S.C., Sec. 2097).

[b]Regulations governing the leasing of these lands are no longer published in the *Code of Federal Regulations*, although statutory authority still exists.

[c]These lands can not be expected to make a significant contribution to U.S. uranium production.

do in order to establish a valid mining claim; this "includes the staking of the claim, sinking a discovery shaft, discovery of a valuable mineral, posting a notice of location, and recording the claim."[10]

Most withdrawals of public-domain land are made for purposes which are inconsistent with uranium mining, although the Bureau of Reclamation has established regulations to receive applications to have reclamation land reopened to mining claims. There are two situations in which withdrawals can involve uranium leasing:

1. Lands withdrawn from entry and location under the general mining laws, for use of the Atomic Energy Commission (AEC) or its successor, the Department of Energy (DOE), are leased by DOE according to 10 C.F.R. 760.1. These regulations were originally issued by the AEC in 1956 and are known as Circular 8.[11] There are two sources of statutory authority for land withdrawals of this type: the Atomic Energy Act of 1954, 42 U.S.C., Sec. 2011, which authorizes DOE to withdraw land, and the Federal Land Policy and Management Act of 1976, 90 Stat. 2743, which authorizes the Secretary of the Interior to withdraw land. The latter replaces the Withdrawal Act of June 25, 1910, 43 U.S.C., Sec. 141–3.[12]

2. Lands withdrawn from entry and location for purposes other than AEC, ERDA, or DOE use can be leased by DOE. Regulations governing this type of uranium leasing were issued by the AEC in 1957, and are known as Circular 9. Although their publication in the *Code of Federal Regulations* was discontinued in 1976, they are still in effect.[13] Statutory authority is found in the Atomic Energy Act of 1954. According to the *Code of Federal Regulations*, these "regulations remain in force and effect insofar as they are not inconsistent with the applicable law governing ERDA [now DOE]."[14]

Other Leasable Acquired Land. There are four categories which can be distinguished:

1. National forest lands are leased by the Bureau of Land Management (Department of the Interior) for the mineral exploration and development, even though the Forest Service (Department of Agriculture) is responsible for managing the surface resources. As table 2–1 indicated, national forests represent nearly half of all acquired land in the United States. The Union Pacific Railroad currently holds an acquired lands mineral lease in Medicine Bow National Forest, Wyoming for the extraction of uranium, vanadium, and associated minerals.[15]

2. Certain other lands can also be leased by the Bureau of Land Management. Various sources of statutory authority are involved; these can be found in the introduction to the regulations, 10 C.F.R., Part 3500. Acquired lands controlled by the Interior Department can be leased on the basis of a broad interpretation of authority conferred by the act of Sep-

tember 1, 1949, 63 Stat. 683.[16] However, as table 3-1 shows, most of these lands are used for purposes previously discussed. The Bureau of Indian Affairs administers Indian reservations. The National Park Service, Bureau of Land Management, and Fish and Wildlife Service administer lands for national park, monument, and wildlife purposes. The only major portion of acquired land held by the Interior Department is Bureau of Reclamation land, which is used for dams and irrigation canals.

3. All other acquired lands can be leased by DOE if they are to be leased for uranium development at all. The Atomic Energy Act of 1954 provides statutory authority, and in the case of lands which are not under DOE control, Circular 9 contains the regulations. There is only one major kind of land which falls under this category, and that is the acreage held by the Department of Defense—nearly, one-quarter of all acquired land.

4. Lands controlled by DOE can be leased by DOE. The Atomic Energy Act of 1954 provides statutory authority, and Circular 8 (10 C.F.R. 760.1) contains the regulations. Circular 8 covers "uranium in public lands withdrawn from entry and location under the general mining laws for use of ERDA and in certain other lands under ERDA control."[17] By definition these "certain other lands" must be acquired lands, since they could not be leased otherwise. At present, 82 percent of DOE's acquired lands are part of the Hanford Reservation, the Savannah River Reservation, and the Idaho National Engineering Laboratory,[18] all of which contain military facilities and high-level radioactive wastes. Land is used as a "buffer" for these operations, so uranium mining on it is unlikely. The remaining 18 percent also involve uses which may conflict with uranium production.

Lands Subject to Uranium Mining Claims

Vacant Public Domain Land. Public land is said to be vacant wherever it "is not reserved, appropriated, or set aside for a specific or designated purpose."[19] Grazing districts are not considered to be land "set aside." A *grazing district* is "an administrative subdivision of the rangelands under jurisdiction of the Bureau of Land Management, established pursuant to Section 1 of the Taylor Grazing Act, to facilitate management of their resources."[20] Valid mining claims on vacant public land can be patented, for both surface and subsurface ownership, without having restrictions on surface use imposed by the patent. The lack of conflict with other kinds of land use (except grazing) makes vacant public land particularly attractive to the mining industry. In a sense, this land consists of the "leftovers" remaining after all the various federal agencies have had a chance to reserve, appropriate, and set aside the land which they can use. Environmental damage on vacant public land has gone unattended[21] and has aroused interest among environmental groups in mining law reform.[22]

Lands Containing Leasing Act Minerals. The Mineral Leasing Act of 1920 established federal authority for leasing of oil and gas, coal, oil shale, phosphate, and sodium on public-domain lands.[23] Subsequent acts (passed in 1926, 1927, and 1932) provided for the leasing of sulfur and potassium. In addition, the act of September 2, 1960 added native asphalt, solid and semisolid bitumen, and bituminous rock. Since this amendment followed the Multiple Mineral Development Act of 1954, these minerals were not originally included among "Leasing Act minerals."[24] Although the Mineral Leasing Act did not explicitly prohibit mining claims (for minerals not subject to leasing) on the lands involved, the Interior Department felt that such claims would be invalid. Moreover, although the Mining Law of 1872 did not explicitly prohibit the reservation of leasable minerals in connection with a mineral patent, the Interior Department felt that such a reservation could not be made.[25] The Multiple Mineral Development Act of 1954[26] resolved this legal uncertainty by validating claims on lands containing Leasing Act minerals (those previously listed)[27] and by including a reservation of Leasing Act minerals in subsequent mineral patents. The regulations in 43 C.F.R., Part 3740 are presently used to implement this legislation.

National Forest Lands. Our national forests have been established largely by public land withdrawals, but the act of June 4, 1897 specifically states that these must not include the withdrawal of land from entry and location under the general mining laws.[28] National forests, like Indian lands and land held for plutonium production facilities, are called *reservations*. A reservation is "a withdrawal, usually of a permanent nature; also, any Federal lands which have been dedicated to a specific purpose."[29] The Forest Service has established regulations for the protection of surface resources (36 C.F.R. 252). Restrictions on exploration and mining activity on the Forest Service's public-domain land are in some ways comparable to the restrictions on its acquired land, even though the public domain is not leased.

Although the Forest Service holds 47.9 percent of all acquired lands, only 14.4 percent of its holdings are on acquired land.[30] This situation creates a lot of friction between the Forest Service and the mining industry.[31]

Lands Sold or Conveyed with the Mineral Rights Reserved to the United States. We shall use the term *locatable minerals* to refer to minerals which are not Leasing Act minerals—in other words, minerals covered by the Mining Law of 1872. Other terms for locatable minerals are *metal-liferous minerals* and *hard minerals*.

There are six sources of statutory authority for the reservation of locatable minerals:

1. The Stockraising Homestead Act of December 29, 1916 provides that "all entries made and patents issued under its provisions shall contain a reservation to the United States of all coal and other minerals in the lands so entered and patented."[32] During the 70 years prior to this act, locatable minerals had not been reserved in lands sold or given away by the federal government. For example, the Homestead Act of 1862 contained no such reservation. After 1916, however, minerals were reserved under several different statutes. Mining activity on lands patented under the Stockraising Homestead Act is regulated according to 43 C.F.R. 3814.

2. Color-of-title acts, beginning with the act of September 21, 1922, provide for the sale of federal lands which have been occupied or held by people who mistakenly believed them to be nonfederal. *Color of title* consists of "any fact, extraneous to the act or mere will of the claimant, which has the appearance, on its face, of supporting his claim of a present title to land, but which, for some defect, in reality falls short of establishing it."[33] Minerals are reserved on these lands according to the regulations in 43 C.F.R. 2541.3.[34]

3. The act of March 20, 1922, as amended, provides for the conveyance[35] of federal lands (or timber) in exchange for nonfederal land which may be added to the national forest system. Minerals may be reserved on the land which is conveyed under these "forest exchanges." In the case of public-domain lands, the minerals are subject to mining claims.[36]

4. The Taylor Grazing Act of June 28, 1934 authorized the Secretary of the Interior to convey public-domain land in exchange for state land or private land. Minerals may be reserved on the land which is patented.[37]

5. The Alaska Public Sale Act of August 30, 1949 provides for the reservation of minerals on certain lands sold in Alaska. These minerals are open to location but cannot be patented. Mining activity is regulated according to 43 C.F.R. 3822.

6. The Federal Land Policy and Management Act of October 21, 1976 includes a very broad requirement regarding the reservation of minerals:

> All conveyances of title issued by the Secretary [of the Interior], except those involving land exchanges provided for in Section 206, shall reserve to the United States all minerals in the lands, together with the right to prospect for, mine, and remove the minerals under applicable law and such regulations as the Secretary may prescribe, except . . . [when the Secretary] finds (1) that there are no known mineral values in the land, or (2) that the reservation of the mineral rights in the United States is

interfering with or precluding appropriate non-mineral development of the land and that such development is a more beneficial use of the land than mineral development.[38]

Lands Withdrawn Without Precluding Claims. It is possible for public-domain land to be withdrawn from certain types of use without being withdrawn from entry and location under the general mining laws. There are three major categories of land subject to such limitations:

1. Stock driveway withdrawals were opened to mining location by the act of January 29, 1929. A mining company seeking to mine these lands does not face any major obstacles related to multiple use, that is, grazing of livestock. There are only two basic restrictions: mining areas must be fenced off to prevent livestock from entering hazardous areas, and watering places must remain accessible to livestock. Regulations in 43 C.F.R. 3815 are applicable.

2. Reclamation withdrawals were partially opened to location by the act of April 23, 1932. A mining company seeking to extract minerals from these lands must file an application with the Bureau of Land Management, "and must set out the facts upon which is based the knowledge or belief that the lands contain valuable mineral deposits."[39] If the Bureau of Reclamation decides that the mining activity described in the application would not interfere with necessary irrigation projects, the lands will be opened. Regulations in 43 C.F.R. 3816 are applicable.

3. Power-site withdrawals were partially opened to location by the act of August 11, 1955. Lands included in any power project which is operating, under construction, or being surveyed (under a preliminary permit) are not open to mining. On other power-site lands, the mining company must bear the risk of having its facilities wiped out at any time to make way for power generation projects. No application is necessary, but no guarantee against federal interference is offered. Regulations in 43 C.F.R., Part 3730 are applicable.

Lands Containing Uranium-Bearing Lignite. Some of the lignites in South Dakota, southwestern North Dakota, and eastern Montana contain significant amounts of uranium—as much as 1.0 percent.[40] Anything containing 0.05 percent or more uranium by weight constitutes "source material."[41] Large-scale recovery of uranium from lignite was seriously considered as early as 1954.[42] However, all lignites had been withdrawn, by the Mineral Leasing Act of 1920, from entry and location under the general mining laws. The act of August 11, 1955 permitted deposits of "valuable" source material to be claimed under the mining laws "unless embraced within a coal prospecting permit or lease."[43] Under this act, all unpatented claims expired on August 11, 1975, except those for which a valid application for patent was filed prior to that date.[44] Mining of uranium-bearing lignites is regulated according to 43 C.F.R., Part 3720.

Other Lands. There are a few areas, subject to special mining laws, which cannot be expected to make a significant contribution to U.S. uranium production. Possibly they will never make any contribution at all. For the sake of completeness, we list them here:

1. The Oregon and California Railroad and Reconveyed Coos Bay Wagon Road Grant Lands, known as O&C lands, are open to location and entry under the act of April 8, 1948. Regulations in 43 C.F.R. 3821 apply.

2. National forest wilderness lands are open to location until midnight December 31, 1983, "subject to the provision of such regulations as may be prescribed by the Secretary of Agriculture pursuant to Section 4(d)(3) of the Wilderness Act."[45] The Bureau of Land Management also administers regulations, which are listed at 43 C.F.R. 3823. The mining industry's view of these regulations is suggested by one attorney's comment that "concern with Mining Law activities within wilderness areas may be largely academic."[46]

3. The portion of the watershed of the city of Prescott, Arizona lying within the Prescott National Forest is open to location and entry, subject to the provisions of the act of January 19, 1933. Regulations in 43 C.F.R. 3824 apply.

4. The Papago Indian Reservation, Arizona is open to location, although it has been closed to entry since May 27, 1955. Regulations in 43 C.F.R. 3825 apply. The act of June 18, 1934 opened these lands to mining claims.

5. Five National Park Service areas—Mount McKinley National Park, Olympic National Park, Death Valley National Monument, Glacier Bay National Monument, and Organ Pipe Cactus National Monument —are open to location. Regulations in 43 C.F.R. 3826 apply.

6. King Range National Conservation Area is open to location, subject to the regulations found in 43 C.F.R. 3827.

Among all these areas, only the O&C lands are free of conflicts between mining and other types of land use.

Relative Contributions to Uranium Production

We have seen that the various categories of federal land shown in table 2–2 are managed under an assortment of regulations. This is not only complicated, but also bewildering as well. Each one of the categories in Table 2–2 involves a certain legal and political context which can be examined in detail. For example, the question of how the government should manage public-domain lands containing both Leasing Act minerals and uranium generated an extensive legal literature in the 1950s. One writer has summarized the debate as follows:

> The story behind this is at once romantic and highly legalistic. The first adjective is used to describe the uranium boom. The last is meant to convey a kaleidoscopic picture of legislation which is so boring that even lawyers shudder when perchance they encounter almost any volume of the proceedings of the Rocky Mountain Mineral Law Institute.[47]

In view of this sort of complexity, it is helpful to get an idea of the relative contributions of different types of federal land to uranium production so that attention may be focused on those which have a major impact. We may begin by looking at past production.

Data on leased Indian lands are maintained by the Bureau of Land Management. Production figures are presented in table 3–3. Data on other leased acquired lands and public-domain lands are available from the same source. Cumulative ore production on these lands through fiscal year 1975 totaled 211,542 tons, or about 1.1 percent of cumulative ore production on Indian lands.[48] All but 10,995 tons (of the 211,542 tons) came from public-domain lands, which could be leased only by the AEC. Taken together, these figures show that federal lands can be ranked by past uranium production as follows:

1. Public-domain land subject to claims
2. Indian reservations, leased by BIA
3. Public-domain land, leased by AEC
4. Acquired land, leased by BLM

The third category is minor, and the fourth insignificant. It would be desirable to break down the first category, but data are unavailable.

The primary reason for the unavailability of data on the relative contribution to uranium production of different types of land subject to claims is that federal records were not required to be kept until the Federal Land Policy and Management Act of 1976 was enacted. Under this legislation, records of unpatented claims located prior to October 21, 1976 need not be filed with the federal government until October 21, 1979.[49] It is difficult to guess how much time will then be required for the government to publish comprehensive data on these and subsequent claims. Some idea of the difficulty of the task may be obtained from an account of the uranium boom of the 1950s published in 1967. At that time, the Interior Department estimated that 6 million claims had been filed under the 1872 mining law and half a million were still active. One

Table 3–3

Uranium Ore Production on Indian Lands, Compared to Total U.S. Production

(*short tons ore*)

Year	Production on Indian Lands	Ore Receipts at Buying Stations and Mills	Percent
1950	36,963	251,000	14.73
1951	44,582	347,000	12.85
1952	66,693	435,000	15.33
1953	140,897	734,000	19.20
1954	219,672	1,106,000	19.86
1955	262,977	1,524,000	17.26
1956	694,313	3,005,000	23.11
1957	1,328,265	3,695,000	35.95
1958	1,593,613	5,178,000	30.78
1959	1,544,178	6,935,000	22.27
1960	1,402,036	7,970,000	17.59
1961	1,486,925	8,041,000	18.49
1962	1,256,946	7,053,000	17.82
1963	831,603	5,948,000	13.98
1964	749,674	5,297,000	14.15
1965	677,556	4,376,000	15.48
1966	464,059	4,329,000	10.72
1967	371,344	5,272,000	7.04
1968	516,393	6,448,000	8.01
1969	557,161	5,904,000	9.44
1970	679,533	6,324,000	10.75
1971	771,799	6,279,000	12.29
1972	724,261	6,418,000	11.28
1973	831,148	6,537,000	12.71
1974	1,136,739	7,027,000	16.18
1975	1,006,398	7,057,000	14.26

Source: Production on Indian Lands is reported in U.S. Department of the Interior, Geological Survey, *Federal and Indian Lands Coal, Phosphate, Potash, Sodium, and Other Mineral Production, Royalty Income, and Related Statistics: Fiscal Year 1975* (Washington: GPO, 1976). Ore receipts are reported in U.S. Energy Research and Development Administration, *Statistical Data of the Uranium Industry*, GJO-100 (76) (Grand Junction, Colo.: USERDA, 1976), p. 10.

observer suspected that a quarter of all these claims were uranium claims: "I estimate that of the six million claims, one and a half million or 25 percent have been filed in the last fifteen years in connection with uranium."[50]

Although the mid-1950s were a time of wild speculation in uranium, the large number of claims is partially attributable to the mining laws. There are two types of claims: lode and placer. The first follows a subsurface ore deposit, not to exceed 20.66 acres in area,[51] and the second follows a surface boundary, containing no more than 20 acres per individual claimant and no more than 160 acres per claim.[52] The area of claim land held for uranium exploration and mining as of January 1, 1977 was 15,067,000 acres, according to an ERDA estimate.[53] If we assume an average of 100 acres per claim, there must be at least 150,670 claims involved. This is only a minimum figure, because it assumes no overlap or duplication. Many claims are partially or wholly invalid because they cover land which has already been claimed. Until fiscal year 1976, the AEC and ERDA obtained data on claims through contracts with Lucius Pitkin, Inc. The Federal Trade Commission reported in 1975 that "the number of [valid] claims abstracted by Pitkin is arbitrarily reduced by 25 percent to account for abandonments and overstaking."[54] However, if 25 percent of the half million claims active in 1967 were filed in connection with uranium, then 125,000 claims were involved. Whatever the true figures are, it is clear that recording problems are inevitable under the present mining laws.

Subsurface Ownership

In connection with lands subject to claims, we have already presented two categories of land in which the federal government maintains subsurface ownership of locatable minerals, that is, owns the mineral rights but not the surface resources: (1) lands which were formerly public-domain lands, but were sold, with the minerals reserved to the United States, and (2) lands which were formerly public-domain lands, but were conveyed in exchange for additions to national forests or grazing lands subject to the Taylor Grazing Act. Surface resources were conveyed, but minerals were reserved to the United States. In order to trace the historical evolution of statutory authority, we lumped these together under one heading (listed in table 3–2).

Logically, one might expect four other types of land to be listed:

3. Lands which were formerly public-domain lands, but were granted to homesteaders, states, railroad corporations, or other parties, with the minerals reserved.
4. Lands which were formerly acquired lands, but were conveyed in exchange for additions to the national forests, with minerals reserved to the United States.[55]
5. Lands which were formerly acquired lands, but were sold, with the minerals reserved.[56]
6. Lands for which the mineral rights alone have been acquired by the government through purchase, condemnation, or gift.

However, we have not found any reference to such lands in published literature, and apparently they are scarce or nonexistent. With regard to the fourth and fifth categories, this scarcity is not surprising; acquired lands are generally held for permanent use and are often located in areas where mining would not represent the most economically efficient use of the land. As far as uranium is concerned, the only rationale for the sixth category would be the unitization of properties subject to leasing. However, most uranium leasing is done on Indian reservations, where this is not a problem. The purchase of complete mineral rights, in connection with other mineral leasing programs, might facilitate unitization but would be a cumbersome way of accomplishing this objective. Negotiations with the nonfederal owners in a "checkerboard" ownership pattern would be simpler.[57]

The absence of lands in the third category represents a significant aspect of past federal land policies. Moreover, lands in the first category represent only a fraction of federal land sales. Very large areas have been given away or sold, ostensibly for purposes completely unrelated to mining, without the reservation of minerals. Table 3–4 summarizes the acquisition and disposition of the public domain. For every acre patented with all minerals reserved, there are twenty-six acres granted or sold with no minerals reserved. Consequently, subsurface ownership does not play a leading role in the present legal system governing federally owned

Table 3–4
Acquisition and Disposition of the Public Domain

	Land Area (acres)	Percentage of Total
Louisiana Purchase (1803)	523,446,400	29.0
Alaska Purchase (1867)	362,516,480	20.1
Mexican Cession (1848)	334,479,360	18.5
State Cessions (1781–1802)	233,415,680	12.9
Oregon Compromise (1846)	180,644,480	10.0
Purchase from Texas (1850)	78,842,880	4.4
Cession from Spain (1819)	43,342,720	2.4
Red River Basin	29,066,880	1.6
Gasden Purchase (1853)	18,961,920	1.1
Original Public Domain	1,804,716,800	100.0
Granted or sold with no minerals reserved (net change)	(1,037,619,989)	57.5
Patented with some minerals reserved	(23,833,752)	1.3
Patented with all minerals reserved	(39,450,002)	2.2
Remaining as of June 30, 1974	703,813,057	39.0

Source: U.S. Department of the Interior, *Public Land Statistics, 1975* (Washington: GPO, 1976), pp. 4, 10, 47.

uranium. To understand how this situation arose, we must briefly examine the history of federal policies regarding locatable minerals on the public domain.

Evolution of Federal Policy toward Locatable Minerals

For the purposes of this discussion, we need only present a brief sketch of historical trends.[58] Five periods can be distinguished:

1. From 1781, when the original thirteen states began to cede land west of the Alleghenies to the federal government,[59] until May 20, 1785, when a land ordinance was enacted by the Continental Congress, there was no federal policy toward minerals on the public domain. If this seems odd, one must remember that there was no Constitution either.

2. From May 20, 1785 until July 11, 1846, when Congress authorized the sale of lead mines in Arkansas and in the upper Mississippi valley, there were limited and sporadic attempts to ensure that lands known to be valuable for minerals would be withdrawn from entry under the agricultural land laws. The Land Ordinance of 1785 stated that "there shall be reserved . . . one-third part of all gold, silver, lead and copper mines, to be sold, or otherwise disposed of as Congress shall hereafter direct."[60] After the Continental Congress dissolved, this ordinance was not reenacted. In 1807 Congress reserved lead mines in the Indiana Territory, and a leasing program for lead mines was begun. However, in 1829 Congress abolished the leasing program in Missouri and authorized the sale of the mines. In 1834 the President directed that mineral lands be reserved from entry in the Wisconsin Territory, but by 1840 these lands had mostly been granted or sold, primarily due to fraud. Finally in 1846 lead leasing was abandoned.

3. From July 11, 1846 to December 29, 1916, when the Stockraising Homestead Act was approved, no attempt was made to retain federal ownership of locatable minerals, except in the case of certain land withdrawals which were inconsistent with the sale of these minerals. National parks, beginning with Yellowstone in 1872, involved restrictions on entry; Indian reservations were maintained; and certain oil lands were withdrawn in 1909 by President Taft. Otherwise, this 70-year period was characterized by vast giveaways of locatable minerals. No attempt was made to reserve mineral rights on lands sold or conveyed by the government. This federal generosity was sustained not only by political corruption, but by the prevailing attitudes toward land: "This was a period in which the country fervently believed that land was an unlimited asset to be disposed of as rapidly as possible in the interest of quick settlement and exploitation in order that all might prosper."[61]

4. From December 29, 1916 to October 21, 1976, when the Federal Land Policy and Management Act was passed, there was in effect a fragmented policy favoring the reservation of minerals on lands sold or conveyed by the government. Various statutes were passed to provide for mineral reservations related to specific types of land, but no overall statement of objectives was enacted.

5. From October 21, 1976 to the present, federal mineral policy has included two key assertions: first, that federal land should not be sold or conveyed in connection with land-use plans, and second, that minerals should normally be reserved whenever a sale or conveyance is made. The opening declaration of the 1976 act states the first objective:

> The Congress declares that it is the policy of the United States that—(1) the public lands be retained in Federal ownership, unless as a result of the land use planning procedure provided for in this Act, it is determined that disposal of a particular parcel will serve the national interest. . . .[62]

We have quoted the section regarding the second objective in the discussion of lands subject to claims.

Throughout this whole history, from 1781 to the present, there have been controversies over the question of how minerals on the public domain should be managed. The heart of the controversy has been the fact that vast mineral resources have been held in government ownership, while popular values and beliefs have favored an economic system based on private ownership. It has been difficult to achieve a consensus on how the government's wealth should be distributed among individuals, states, private corporations, and so forth. A series of minerals have commanded the public's attention—lead, gold, coal, oil, uranium, outer-continental-shelf oil—while the same fundamental conflict continues to be debated: "The history of public mining law in this country has yet to record a plateau of comparative quietude."[63]

Today the laws governing uranium resources are practically identical with those governing other metals—copper, lead, zinc, and so forth. Among energy resources, however, uranium occupies a unique position. Until 1920, oil and gas were subject to the Mining Law of 1872. "Now, however, only the disposal of uranium is still conducted under its provisions, with most other resources covered by various special mineral leasing statutes."[64] Not surprisingly, there now exist proposals to bring uranium entirely under leasing programs. Opposition to these proposals is based primarily on beliefs which were expressed by Senator King of Utah in 1919 with regard to proposals for coal, oil, and gas leasing: "I am absolutely opposed to the leasing system, the paternalism, the bureaucracy, the autocracy, the un-American system that the leasing system entails. . . ."[65]

In addition, opposition is influenced by regional conflicts, also expressed in 1919:

> Senator Thomas of Colorado expressed the resentment of Westerners, great chunks of whose states "never can pass into private ownership" or contribute tax revenues to the States, at Easterners "living in states unburdened by such conditions, states which were once public-land states also (*sic*), but whose lands have passed entirely into private ownership."[66]

In fact, the seventeen original Eastern states (including Kentucky, but not Florida) were never public-land states. Moreover, the federal government ceded all of Tennessee to the state of Tennessee, allowed all of Texas to be held by the state of Texas, and held very little public domain land in Hawaii.[67] The real discrimination lies between the twelve public-land states west of the hundredth meridian (the Far West and Alaska) and the eighteen public-land states east of the hundredth meridian (the Midwest). The federal government owns 63.6 percent of the former and 4.6 percent of the latter.[68] The current political debate over the mining laws is strongly influenced by this ownership pattern.

Acreage Limitations on Leases

There is another issue which plays an important role in generating opposition to leasing proposals, but is not introduced in congressional debates with the vitriolic fervor characteristic of the antigovernment and anti-Eastern themes. The simple fact is that the general mining laws impose no restrictions at all on mineral exploration activity on the public domain, except to protect one miner or mining company from infringement by another. Restrictions exist only as a result of laws governing the use of surface resources for purposes other than mining, or as a result of environmental legislation. For example, mineral exploration in national forests is subject to Forest Service regulations, and the use of off-the-road vehicles (ORVs) is restricted by Bureau of Land Management regulations. With these exceptions, the use of public-domain land for exploration is not subject to acreage limitations and does not even require notification of the federal government.[69] Leasing, of course, does involve acreage limitations and may involve other restrictions on exploration activity.

As noted in chapter 2, a single company's detailed reconnaissance program, including wide-spaced drilling, may involve areas up to a few hundred square miles. In the past, federal uranium leases have not accommodated this sort of program. As table 3–5 shows, the average area per lease held in 1974 was on the order of 1 square mile (640 acres). Prospecting permits for locatable minerals on acquired lands may not

Table 3–5
Average Acreage per Lease, for Different Categories of Federal Uranium Leases, as of June 30, 1974

Type of Land	Leasing Agency	Number of Leases	Total Acreage	Average Acreage per Lease
Indian reservations	BIA	380[a]	254,380[a]	669
Public-domain lands withdrawn by AEC	AEC	43[a]	25,000[b]	581
Acquired lands	BLM	4[a]	3,808[a]	952

[a]U.S. Federal Trade Commission staff, *Staff Report on Mineral Leasing on Indian Lands* (Washington: GPO, 1976), p. 8.

[b]"First Leases on AEC Uravan Lands Up for Bidding," *Nuclear Industry*, August 1973, pp. 34–35. Bids were due April 1, 1974 for eight tracts, May 1 for eighteen additional tracts, and July 1 for the last seventeen tracts. Some leases, therefore, had not been signed by June 30, 1974, although the acreages were known because they were set by the AEC.

include more than 4 square miles, and a company's total holdings under prospecting permits and leases is normally limited to 32 square miles.[70] By contrast, a company is allowed to lease very large tracts of land for oil and gas exploration; the limitation is 384.5 square miles on public-domain lands plus another 384.5 square miles on acquired lands, per state (slightly more is allowed in Alaska).

Prospecting permits for uranium on lands leased by ERDA under Circular 9 would be limited to 3 square miles. Since Circular 9 is not in use, this limitation is not very significant.[71]

Four different types of acreage limitations in uranium leasing are shown in table 3–6. First, there is an acquired land lease issued by the Bureau of Land Management, for which the acreage is 4 square miles— the maximum which would be allowed for a prospecting permit. Second, there is an Indian land lease in which no exploration activity is involved, and only 2 square miles are necessary for mining and milling. The third lease also involves Indian lands, but a very large area is opened for exploration in the first year, and only about one-seventh of this remains open in subsequent years. The fourth lease is similar to the third, for less than one-eighth of the original exploration area can be leased for mining and milling purposes.

It is clear that the Mobil Oil and Exxon leases offer these companies the opportunity to conduct wideranging reconnaissance programs which would not be possible within conventional uranium leases. The arrangement may also help to resolve a chronic difficulty with uranium leasing on Indian lands: 64.4 percent of the 2,875 tracts offered at competitive sales prior to 1974 were not bid on at all.[72] Presumably the transactions cost of

Table 3–6
Acreage Involved in Four Selected Uranium Leases

Date of Lease	Location	Company	Activity	Acreage
1970	Medicine Bow National Forest, Wyoming	Union Pacific Railroad	Exploration and mining	2,560
1976	Spokane Reservation, Washington[a]	Phelps Dodge	Mining	552
			Milling	670
				1,222
1975	Ute Mountain Reservation, Colorado	Mobil Oil	Exploration:	
			1st year	162,176
			later years (est.)	23,040
1977	Navajo Reservation	Exxon	Exploration:	
			1st & 2nd years	400,000
			3rd & 4th years	200,000
			Later years	150,000
			Mining and milling (maximum acreage)	51,200

Source: U.S. Nuclear Regulatory Commission, U.S. Department of the Interior, and U.S. Department of Agriculture, *Final Environmental Statement Related to Operation of Bear Creek Project, Rocky Mountain Energy Company*, NUREG-0129 (Washington: USNRC, 1977), pp. 1-4, C-1; U.S. Department of the Interior, Bureau of Indian Affairs, *Final Environmental Impact Statement, Sherwood Uranium Project, Spokane Indian Reservation*, FES 76-45 (Portland, Ore.: USBIA, 1976), pp. 1–6, 2-138; U.S. Department of the Interior, Bureau of Indian Affairs, *Final Environmental Statement of the Approval by the Department of the Interior of a Lease of the Ute Mountain Tribal Lands for Uranium Exploration and Possible Mining*, FES 75-94 (Albuquerque, N.M.: USBIA, 1975), pp. 1–2; U.S. Department of the Interior, Bureau of Indian Affairs, *Final Environmental Statement, Navajo-Exxon Uranium Development*, DES 76-60 (Billings, Mont.: USBIA, 1976), pp. 1-4 to 1-6.
[a]The Spokane Reservation contains 154,600 acres.

signing hundreds of small leases prevented companies from prospecting on lands which could easily be explored under a lease of the Navajo-Exxon variety. Similarly, the attractiveness of proposed leasing programs for the public domain would be substantially improved by the availability of much larger acreages leased for exploration purposes than for mining and milling purposes.

Royalties on Leases

Until 1975, average royalties on leased Indian lands remained in the range of 10 to 12 percent of ore values. A minimum royalty rate of 10 percent is required by BIA regulations (25 C.F.R. 171.15). Available data are shown in table 3–7. The low of $1.96 per ton in 1974 may be compared with a high of $8.58 per ton in 1952, where the latter figure is measured in 1974 dollars based on an overall GNP deflator. The implicit price deflator for federal

Table 3–7
Royalties from Uranium Mined on Federal Lands
(*values in current dollars*)

Year(s)	Indian Lands $/ton Ore	Indian Lands Percent of Value	Acquired Lands $/ton Ore	Acquired Lands Percent of Value	Public Domain $/ton Ore	Public Domain Percent of Value
Prior to 1950	2.07	10.01				
1950	1.81	10.00				
1951	3.42	10.02				
1952	4.41	10.00				
1953	3.32	10.55				
1954	3.82	11.15				
1955	3.59	12.44				
1956	2.44	11.58				
1957	2.20	11.67	1.76	12.09		
1958	1.65	11.22				
1959	1.60	11.15				
1960	1.74	11.45				
1961	1.71	11.65				
1962	1.79	11.40				
1963	2.04	11.65	.12	2.68	2.23	6.39
1964	2.07	11.73	.09	2.50	2.99	6.99
1965	2.29	11.80	.68	8.35	2.24	6.46
1966	2.59	11.72			2.82	6.48
1967	2.56	11.74	.12	2.51	2.22	7.34
1968	2.48	11.96			1.00	5.72
1969	2.64	12.02			1.85	6.15
1970	2.67	11.82			1.27	5.94
1971	2.82	11.84			1.27	5.94
1972	3.04	11.99				
1973	2.77	12.24				
1974	1.96	10.10				
1975	2.65	15.73				
Average through 1975	2.21	11.73	.25	5.01	2.43	6.63

Source: U.S. Department of the Interior, Geological Survey, *Federal and Indian Lands Coal, Phosphate, Potash, Sodium, and Other Mineral Production, Royalty Income, and Related Statistics: Fiscal Year 1975* (Washington: GPO, 1976).

government purchases of goods and services yields a royalty of $11.22 per ton (in 1974 dollars) in 1952.[73] Royalties have been lower on acquired lands and public lands, but this comparison is not very significant because only small quantities of ore have been produced under leases on these lands relative to production on Indian lands.

Cash-plus-royalty bidding has been involved in some of the leases issued by the AEC on public-domain lands withdrawn from entry in order to allow the AEC to conduct exploration programs.[74] However, Circular 8 was originally based on cash bonus bidding:

> Except under special circumstances as provided in this section a lease will be issued only to the acceptable bidder offering the highest cash bonus. . . . Under special circumstances, where the Commission believes it is to the best interest of the Government or where the use of competitive bidding may be impracticable, the Commission at its discretion may award or extend leases on the basis of negotiation.[75]

In June 1970 the director of the AEC's division of raw materials met with representatives of the Western Uranium Association (WUA) to discuss this leasing program. The WUA proposed that "the cash bonus . . . be eliminated, substituting either a lottery or a declining royalty bonus with a base royalty to be a percentage of finished mill product," and that "any company or individual be restricted from obtaining more than one lease." [76] These proposals reflected the desire of small mining companies to prevent Union Carbide, American Metal Climax, Atlas Corporation, and other large companies from taking all the leases. The present regulations reflect a compromise reached by the AEC:

> The bid may be on a cash bonus, royalty bonus, or other basis as specified in this invitation to bid. Invitations to bid on some of the lands may be limited to small business concerns as defined by the Small Business Administration, and such invitations may limit the number of leases to be awarded to each bidder.[77]

Among the forty-three tracts leased by the AEC in 1974, eighteen were set aside for small businesses and offered on a deferred cash bonus basis, with no requirement for payments in advance of production.[78]

Among the AEC tracts open to royalty bidding, rates as high as 33.51 percent were offered by the winning bidders. In all, thirty-nine of the forty-three tracts, or 90.7 percent, were bid on by four or more bidders, and all the tracts had at least two bidders. Among tracts offered at uranium sales on Indian reservations through 1974, only 2.3 percent had four or more bidders, 20.6 percent had only one bidder, and 64.4 percent had no bidders at all.[79] This contrast between AEC and Indian land leasing is due not only to the differences in previous exploration work, timing, and bidding procedures, but also to the presence of vanadium in the AEC's ore deposits. For every pound of uranium there was about $8 worth of vanadium in 1970: "At something over $1.60 a pound and at a ratio of five pounds of vanadium to one of uranium in the ore, it is debatable whether Uravan Mineral Belt ore is uranium or vanadium ore." [80] The AEC paid only 31 cents a pound for vanadium oxide in connection with its uranium ore purchasing program; no payment was made for vanadium in excess of 10 pounds of V_2O_5 per pound of U_3O_8, except by special agreement. The price increase made Uravan ores very attactive.[81]

Origins of the Present Mining Law

The basic statute governing the location-patent system is the Mining Law of May 10, 1872, which modified and added to the provisions of the Mining Act of 1866 and the Placer Act of 1870. These laws, in turn, codified the basic rules which had gained acceptance among miners since the California gold rush. The principle of "first come, first serve" was applied by members of the mining camps in order to settle disputes over discoveries of valuable deposits. The rules of these camps were "backed up primarily by a hemp rope" and enforced by the miners themselves.[82] Prior to the Mining Act of 1866, the federal government had not granted anyone the right to claim gold or any other mineral on the public domain, and mining was simply a form of theft. For the miner, the 1866 act represented an "escape from entire confiscation," as one contemporary observer noted.[83]

If the provisions of the Mining Law of 1872 seem unnecessarily cumbersome—for example, if the rule that an individual cannot locate a placer claim larger than 20 acres seems odd—one need only think of El Dorado and the hemp rope, and it all makes sense. If the application of these laws to uranium seems curious, one must recall a certain similarity with gold: "The uranium boom, which reached its height in 1954, was not unlike the California gold rush except that the average prospector was generally not quite as undernourished."[84]

Nevertheless, there are a number of difficulties (which we shall present shortly) involved in the application of nineteenth-century statutes to modern industry. The basic reason why the Mining Law of 1872 has proved to be remarkably resistant to change, despite the fact that it has been known to be "nearly absurd when applied to the discovery and extraction of uranium,"[85] is that there has been no consensus about what should replace it. Most of the critics of the Mining Law of 1872 have apparently preferred to live with it rather than to compromise with their political opponents in developing new legislation.

Difficulties with the Mining Law of 1872

Broadly speaking, there are three areas in which the Mining Law of 1872 presents problems:

1. Conflicts among miners
2. Conflicts between miners and the general public
3. Conflicts between miners and other users of public-domain lands, including federal agencies

The third area must be examined with reference to specific categories of land. The general nature of such conflicts should be clear from our discussion of the various lands subject to claims. Only the first and second topics will be examined here.

The claims system was originally developed in order to resolve conflicts among miners, so these disputes cannot be considered an area of neglect. The basic problem is that technological change has rendered the old rules inadequate. There are three principal difficulties:

1. *Definition of discovery.* The Mining Law of 1872 states that a claim is valid if the claimant has made a "discovery" of a "valuable mineral deposit." The law does not define these terms,[86] and it does not even make clear who is supposed to define them. The first section of the law is basically a legitimization of the "local custom" of taking minerals from the public domain without compensating the federal government:

> Except as otherwise provided, all valuable mineral deposits in lands belonging to the United States, both surveyed and unsurveyed, shall be free and open to exploration and purchase, and the lands in which they are found to occupation and purchase, by citizens of the United States and those who have declared their intention to become such, under regulations prescribed by law, and according to the local customs or rules of miners in the several mining districts so far as the same are applicable and not inconsistent with the laws of the United States.[87]

The 1866 act referred only to "mineral lands," not "valuable mineral deposits," so the question of value was clearly important to the drafters of the 1872 law.[88]

There are three different interpretations of "discovery" corresponding to three different situations. First, known lodes or veins remain open to lode claims, even after patented placer claims are made on the land area involved. It is possible to announce "discovery" of a lode which was known to be valuable at the time the placer claim was patented. The courts have interpreted this sort of "discovery" (a misnomer really) and established very strict requirements for showing that the lode is "valuable." At the time when the application for patent is filed, the existence of the lode must "render the land more valuable on that account, under then existing circumstances, than for any other purpose."[89] This rule applies not only to patents for placer claims, but to patents for town sites, which are not mining claims at all.

Second, one miner may locate a claim, and a second miner may subsequently locate a claim on the same land, stating that the first claim was invalid because the claimant did not really discover a mineral deposit. The courts have interpreted discovery in this context to mean simply proof that the mineral deposit is present, regardless of whether or not it is valuable: "When the locator finds rock in place containing mineral, he has made a discovery within the meaning of the statute, whether the earth or rock is rich or poor, whether it assays high or low."[90] Carried to its logical extreme, this 1893 decision would allow uranium to be "discovered" practically anywhere.

Third, a miner may apply to the Department of the Interior for a patent on a mining claim. The Department interpreted "discovery" in such a context, with the approval of the Supreme Court, in 1894: "Where minerals have been found and the evidence is of such a character that a person of ordinary prudence would be justified in the further expenditure of his labor and means, with a reasonable prospect of success, in developing a valuable mine, the requirements of the statute have been met."[91] There was a time during the 1950s when it appeared inconceivable that a uranium miner could be "a person of ordinary prudence." However, this interpretation has survived. Only a small fraction of uranium claims have ever been patented.

The most important situation with respect to uranium mining is the second one, the case of conflicting claims. According to the Federal Trade Commission, "claims may be as many as seven or eight deep in the case of promising uranium properties."[92] During the 1950s, it became necessary to develop a new standard of proof of the existence of a mineral deposit. The traditional method was to dig out an ore sample and have it assayed: "The only type of exploration contemplated by the mining laws of 1872 was strictly the pick and burro variety."[93]

For the early uranium discoveries on the Colorado Plateau, this situation presented no difficulty because outcrops were found on the walls of mesas. When it became necessary to mine the interior of mesas, under hundreds of feet of overburden, uranium deposits were identified by lowering radiometric instruments into drill holes. Court cases resulted. "In the first case, the Colorado Supreme Court said, in effect, 'Don't try to put this radiometric detection business over on us.' "[94] However, by the 1960s, the courts accepted geophysical evidence of discovery.

Since the AEC refused to purchase ores containing less than 0.1 percent uranium,[95] the disputes of the 1950s must not have involved low-grade deposits. Conceivably the future price of uranium concentrate may reach levels which would justify the cost of filing claims on which the geophysical or geochemical evidence approximates the characteristics of a large portion of the Colorado Plateau. In such a situation, the Mining Law of 1872 would be of little help in resolving disputes.

2. *Exploration rights prior to discovery*. During the gold rush, the sight of a prospector looking for gold in a particular area was not regarded as evidence that he had reason to believe that the area was especially promising. Not only was the prospector's judgment occasionally impaired by visions of El Dorado, but his capital investment consisted of a pick and burro, which did not require that anyone else's judgment be involved in decisions about where to explore. Consequently, it is not surprising that the Mining Law of 1872 does not prevent one prospector from working near another prior to discovery. When the oil industry grew up on the public domain, this presented problems.[96] A drill rig represented a major capital investment for which specific locations had to be chosen carefully. Speculators in mineral rights could look for drill rigs and surround them with mining claims, making it difficult for the person who discovered the oil to gain the full rewards of his discovery. This difficulty was resolved by the Mineral Leasing Act of 1920.

Uranium exploration also involves drill rigs, of course. The prospector's only protection lies in a rule of property law which has been applied to mining law. It states that an occupant of a piece of property cannot be denied ownership rights by someone who has no better rights than the occupant. In the context of mining claims, the courts have held that a prospector can exclude others from his claim, or at least the immediate area being explored, as long as he maintains "actual and continuous possession and diligent prosecution of work toward discovery."[97] This phrase is sufficiently vague that it provides little protection to the uranium miner.

3. *Preservation of old rights under new legislation*. Every time a proposal for reform of the mining laws is examined, the question of how to preserve existing rights arises. For example, the leasing proposal submitted by the General Accounting Office in May 1975 contains the following suggestion:

> We further recommend that the Congress . . . Require that, to preserve valid existing rights, mining claims be recorded with the Department of the Interior within a reasonable period of time after the legislation is enacted and that claimants perfect their claims, before their claims are recorded, by furnishing evidence that they have made a discovery of valuable minerals.[98]

This is easier said than done. In the case of oil shale, for example, the Interior Department has been trying to determine who has "valid existing rights" ever since 1920. The simplest solution, logically, is for the government to revoke those rights, but this has not been a simple solution politically.

In summary, there are three areas in which the Mining Law of 1872 makes it difficult to delineate the rights of one miner versus another: the definition of discovery, exploration rights prior to discovery, and preservation of old rights under new legislation.

We may now turn our attention to conflicts between mining interests and the public interest. Again, there are three major deficiencies:

1. *Nonexistent or inadequate payment for mineral rights.* The Mining Law of 1872 allows miners to extract valuable minerals on unpatented claims on federally owned land without paying the government for the resources removed. The inequitable nature of this arrangement was obvious from the very beginning. The Comstock lode, which was discovered in 1859, produced $50 million by 1865 and nearly $100 million by 1868. The Mining Act of 1866 "made it possible for the Comstockers to secure ownership from the federal government without charge."[99]

On patented claims, the government does receive payment, but only a modest amount—$2.50 an acre for placer claims and $5.00 an acre for lode claims. The General Accounting Office has found that the resale of patented claims can be quite profitable. "For example, 150 acres of land patented in California in August 1959 for $375 were sold within 15 months for $43,500. Another 80 acres of land patented in Arizona in 1955 for $200 were sold in 1972 for $368,000."[100] Once a claim is patented, the land need not be used for mining purposes. The GAO found that patented claims in Arizona had been used for residential development and for expanding a university complex.[101] However, it is illegal to use unpatented claims for purposes other than mining.

2. *Absence of environmental controls.* The Mining Law of 1872 makes no provision for land reclamation or any sort of environmental regulation. The historical reasons for this omission are obvious. First, mining activity in 1872 (other than coal mining, which never came under the mining law) was on a smaller scale than it is today and did not involve the movement of massive amounts of overburden and waste material. Second, pollution of the environment was not a major concern of Congress in the 1870s. What is remarkable is that no laws have been passed to regulate mining law activities since that time, except with regard to specific categories of land subject to nonmining use (for example, national forests).

Uranium mining and milling involve the release of radioactive materials into the air, water, and soil. Consequently, the environmental problems involved are not typical of the mining and processing of other minerals. We shall discuss the regulation of uranium production in a later section of this chapter.

3. *Absence of federal control over production rate.* The Mining Law of 1872 does not provide the government with any methods of speeding up or slowing down the production of locatable minerals. The controls which are available originate in other laws which do not involve mining directly.

Two methods of regulating production exist: one is to withdraw lands from location and entry under statutes covering types of land use which are inconsistent with the mining of locatable minerals, and the other is to regulate surface use of the public domain under statutes covering activities which are consistent with such mining. Both these methods are restrictive. Neither patented nor unpatented claims require production or impose cost penalties where production is nonexistent.

In theory, a third method of influencing extraction rates exists: to keep lands open to mining location, but to withdraw them from uses which conflict with mining. However, there appears to be only one situation where such action was taken: 9,787 acres of the Warren mining district in Bisbee, Arizona were withdrawn from nonmineral entry in 1912 in order to allow the development of deep copper mines.[102]

Special AEC, ERDA, and DOE Powers

In 1946 there was a shortage of domestic uranium reserves available to meet military requirements. In January 1947 the government estimated domestic reserves to be 2,000 tons of uranium oxide;[103] the January 1977 estimate, by contrast, was 840,000 tons, plus an additional 140,000 tons from byproduct sources.[104] In view of this shortage, the Atomic Energy Commission was given special powers when it was created under the Atomic Energy Act of 1946:

> The Commission is authorized and directed to purchase, take, requisition, condemn, or otherwise acquire, supplies of source materials or any interest in real property containing deposits of source materials to the extent it deems necessary to effectuate the provisions of this Act. . . . Just compensation shall be made for any property taken, requisitioned, or condemned under this paragraph.[105]

The wording of this section was modified slightly by Public Law 585, 79th Congress, which was enacted on August 13, 1954. Within 3 weeks the wording was changed again, under Section 66 of the Atomic Energy Act of 1954:

> The Commission is authorized and directed to the extent it deems necessary to effectuate the provisions of this Act. (a) To purchase, take, requisition, condemn, or otherwise acquire supplies of source material; (b) to purchase, condemn, or otherwise acquire any interest in real property containing deposits of source materials; and (c) to purchase, condemn, or otherwise acquire rights to enter upon any real property deemed by the Commission to have possibilities of containing deposits of source material in order to conduct prospecting and exploratory opera-

tions for such deposits. . . . Just compensation shall be made for any
right, property, or interest in property taken, requisitioned, condemned,
or otherwise acquired under this section.[106]

The Energy Reorganization Act of 1974 transferred "all the licensing
and related regulatory functions of the Atomic Energy Commission" to
the Nuclear Regulatory Commission (NRC) and transferred all the other
functions of the AEC to ERDA.[107] Therefore, DOE now has the authority
to assert ownership of all uranium ore in the United States, regardless of
the Mining Law of 1872 or practically any other law. The AEC never used
these powers, because it took care to maintain good relations with the
mining industry,[108] as did the Joint Committee on Atomic Energy.

Although the joint committee has been abolished, DOE continues to
uphold this policy. Moreover, the use of DOE's special powers to take
uranium ore could result in the removal of Section 66 from the Atomic
Energy Act. There can be no question that the powers were not established with the commercial nuclear industry in mind. It is reasonable to
assume, therefore, that they will not play a role in future uranium leasing
policies.

Environmental Regulation of Uranium Mining and Milling

At present the environmental aspects of uranium mining and milling are a
controversial matter, although in the U.S. debate on nuclear power, they
tend to receive less attention than some of the other parts of the fuel
cycle. As chapter 2 indicated, there are several different federal agencies
involved in regulating domestic uranium production. Among them, only
the Mine Safety and Health Administration (MSHA) deals with matters
outside the scope of "environmental" issues, as they are commonly
defined. In the following discussion, we will try to clarify a procedural
topic—the question of when an environmental impact statement (EIS)
must be filed—without reviewing all the substantive issues being debated
today. The reason for our focus upon the EIS is that it can have an
important impact on leasing schedules, and on production costs.

There are three kinds of agencies which may issue an EIS on uranium
mills and mines. First, the Nuclear Regulatory Commission will issue an
EIS with regard to an application for a license for a newly constructed
uranium mill, and it may issue an EIS with regard to an application to
renew a license for an existing mill.[109] Second, a state agency may issue an
EIS, as defined by state law, with regard to some action by the appropriate state government. The particular requirements vary from state to
state. And third, a federal agency which leases uranium lands may issue

an EIS with regard to an application for a lease or prospecting permit. Obviously there is a potential for duplication of effort. Unfortunately, there are no clear statements of policy regarding the second and third situations, with specific reference to uranium supply. The policy regarding the first situation is clear, but it is under attack in the courts.

Uranium Mill Licensing

The Nuclear Regulatory Commission issues licenses for uranium mills, but not for mines. The Atomic Energy Act of 1954, as amended, states that "licenses shall not be required for quantities of source material which, in the opinion of the Commission, are important." NRC regulations regarding "unimportant quantities of source material" [10 C.F.R. 40.4(h), 40.4(k), and 40.13(b)] include a specific reference to unprocessed ore:

> "Source material" means (1) uranium or thorium, or any combination thereof, in any physical or chemical form or (2) ores which contain by weight one-twentieth of one percent (0.05 percent) or more of (i) uranium, (ii) thorium or (iii) any combination thereof. . . .
>
> "Unrefined and unprocessed ore" means ore in its natural form prior to any processing such as grinding, roasting or beneficiating, or refining. . . .
>
> Any person is exempt from the regulations in this part [10 C.F.R., Part 40] and from the requirements for a license set forth in Section 62 of the [Atomic Energy] act to the extent that such person receives, possesses, uses, transfers, or imports into the United States unrefined and unprocessed ore containing source material; provided, that, except as authorized in a specific license, such person shall not refine or process such ore. The exemption contained in this paragraph shall not be deemed to authorize the export of source material.

If private industry were to propose construction of uranium mills to process ores containing less than 0.05 percent uranium, the NRC would undoubtedly change these regulations in order to assert licensing authority over such mills. Of course, the NRC also licenses nuclear power plants and other fuel cycle facilities.

The Nuclear Regulatory Commission does not license all uranium mills. Under Section 274 of the Atomic Energy Act of 1954, "the Commission is authorized to enter into agreements with the Governor of any State providing for discontinuance of the regulatory authority of the Commission," subject to the condition that the Commission "may terminate or suspend its agreement with the State and reassert the licensing and regulatory authority vested in it under this Act, if the Commission finds that such termination or suspension is required to protect the public health and safety."[110] This is a curious legal device in which the word *authority* takes on a meaning quite different from everyday usage. The NRC always has the ability to override any licensing decision by a state government; in this sense it is the real "authority." However, it does not have to issue certain kinds of licenses if this licensing activity can be done by state governments without endangering the public health and safety.

Table 3-8 lists the licensing agencies for the eleven states which are most likely to play a role in future uranium production (excluding uranium recovered as a byproduct from phosphate or copper mining). The number of mills licensed by NRC works out to be roughly equal to the number licensed by states, under present conditions.

State Environmental Impact Statements

There are only three states which have enacted legislation requiring comprehensive EIS review. Table 3-9 summarizes the legal picture, taking into account the NRC's environmental statements. The outlook for Arizona, Colorado, Nevada, New Mexico, and Texas is unclear because of a lawsuit filed in federal district court in New Mexico on May 3, 1977. In this case, the Natural Resources Defense Council (a national environmental group), the Central Clearing House (a local environmental group), and two New Mexico residents have sued the NRC, three of its Commissioners, the New Mexico agency which issues uranium mill licenses, and the director of that agency for failing to prepare an EIS on the pending application for United Nuclear's mill at Church Rock.[111] Two observations on this case may be made. First, the National Environmental Policy Act of 1969 (NEPA) requires that federal agencies "prepare detailed environmental statements on proposals for legislation and other major Federal actions significantly affecting the quality of the human environment."[112] The licensing of the Church Rock mill by the state of New Mexico would not be a major federal action. However, the NRC's signing of an agreement with New Mexico on April 3, 1974 could be considered a major federal action. It would appear that an EIS on all the nuclear regulatory functions of the state is required.

Second, regardless of whether or not an EIS is prepared, the substantive matters discussed in the suit regarding the "public health hazards from uranium mill tailings" suggest that the states' current licensing procedures and standards are inadequate to protect the public health and safety. If this is true, the Atomic Energy Act of 1954 requires that the NRC must terminate the agreement, regardless of whether or not the state wants to terminate the agreement.

Table 3–8
Information on Agencies with Authority to License Uranium Mills, and on Production Centers, by State

State	Agency[a]	Effective Date of Agreement[a]	Past Production Center(s) Yielding over 1000 ST U_3O_8[b]	Operating Production Center(s)[c]	Future Production Center(s) Justified by Potential Resources Only[c]
Alaska	U.S. Nuclear Regulatory Commission	—	X		
Arizona	Atomic Energy Commission	May 15, 1967			X
California	Department of Health	Sept. 1, 1962	X		
Colorado	Department of Public Health	Feb. 1, 1968	X	X	
Nevada	Department of Human Resources	July 1, 1972			X
New Mexico	Environmental Improvement Agency	May 1, 1974	X	X	
South Dakota	U.S. Nuclear Regulatory Commission		X		X
Texas	Department of Health Resources	March 1, 1963	X	X	
Utah	U.S. Nuclear Regulatory Commission	—	X	X	
Washington	Department of Social and Health Services	Dec. 31, 1966	X	X	
Wyoming	U.S. Nuclear Regulatory Commission	—	X	X	

[a]All the agencies listed are state agencies, except the U.S. Nuclear Regulatory Commission. This information was provided by the U.S. Nuclear Regulatory Commission in private correspondence.

[b]U.S. Energy Research and Development Administration, National Uranium Resource Evaluation, Preliminary Report, GJO-111(76) (Grand Junction, Colo.: USERDA, 1976), p. 13.

[c]John Klemenic, "Analysis and Trends in Uranium Supply," Uranium Industry Seminar, GJO-108(76) (Grand Junction, Colo.: USERDA, 1976), pp. 229–231.

Table 3–9
Agencies, Other than Federal Leasing Agencies, Required to Prepare Environmental Impact Statements Related to Federal Uranium Leasing

State(s)	Agencies Involved if the Lease Includes Uranium Milling	Agencies Involved if the Lease Excludes Uranium Milling
Arizona, Colorado, Nevada, New Mexico, Texas[a]	None (unless required by pending court case)	None
California	California Energy Resources Conservation and Development Commission (possibly)	California Energy Resources Conservation and Development Commission (possibly)
Washington	Washington Department of Ecology (possibly)	Washington Department of Ecology (possibly)
South Dakota	U.S. Nuclear Regulatory Commission and South Dakota Department of Environmental Protection	South Dakota Department of Environmental Protection (possibly)
Alaska, Utah, Wyoming	U.S. Nuclear Regulatory Commission	None

Source: Information provided by U.S. Nuclear Regulatory Commission in private correspondence.

[a]Texas has no public-domain lands.

The plaintiffs in the suit have not focused on these two issues, so they may lose the case. However, unless they run out of money, they are likely to bring suit on these issues eventually. This is the reason why the controversy introduces a great deal of uncertainty into the production schedules and plans of the uranium mining and milling industry.

Federal Leasing Practice

The National Environmental Policy Act does not provide agencies with much help in determining what constitutes a major federal action "significantly affecting the quality of the human environment." Agencies have developed their own interpretations, and the federal courts have been the ultimate arbiters. Unfortunately, the record of uranium leasing to date is not very informative. Two kinds of leasing actions which require an EIS are suggested: (1) a lease sale held by DOE, involving an area of at least 25,000 acres, and (2) a lease of Indian lands, involving either ore production of at least 7,950,000 tons or an area exceeding 165,000 acres. These situations are not likely to be typical of future uranium leasing. Moreover, the acquired lands lease awarded to the Union Pacific Railroad did not involve an EIS until a licensing application for a uranium mill on nearby land was filed with the NRC. The lease of Lake Mead National Recreation Area has required only an environmental "assessment," not an impact statement.

It is very doubtful that anything done by the Bureau of Land Management in connection with the Mining Law of 1872 could be construed as a major federal action. Thus it is not surprising that environmental groups are generally in favor of reforming or repealing the Mining Law of 1872.

Problems with Federal Environmental Regulation

Since the focus of this book is federal policy, we need not consider deficiencies in state regulation, except insofar as they affect NRC's regulatory activity. At the federal level, several areas of difficulty can be identified. They are problems in the sense that they provide a basis for concern that existing procedures need to be changed, and these concerns will be expressed in legislative proposals and in litigation. There are five areas involved.

1. No federal permits or licenses, and therefore no environmental reviews, are required specifically for the construction or operation of uranium mines unless a particular lease requires such a review.

2. No environmental impact statements are required for uranium mill

licenses in agreement states unless a particular state (California, South Dakota, or Washington) interprets its own laws so as to require one.

3. No construction permit is required for uranium mills in nonagreement states under NRC authority. Therefore, an environmental impact statement may not be prepared until it is too late to consider whether or not a mill should be built in a given location.

4. No environmental review is required for uranium mining claims unless surface use of the land for nonmining purposes involves some major federal action.

5. No clear statement of policy exists regarding the decision whether or not to prepare an environmental statement for a uranium prospecting permit or lease.

It should be noted, also, that responsibility for environmental protection is distributed among many different agencies, and problems may arise in coordinating their work. Three different types of federal agencies may be involved in a single uranium lease: the agency administering mineral resources, the agency administering surface resources, and the regulatory agencies dealing with mine safety, public safety, and environmental quality.

Proposals for Changing the Present Legal System

We have seen in this chapter that there are difficulties with both the Mining Law of 1872 and with federal environmental regulation as they pertain to uranium resources. To conclude the discussion let us examine some proposals for change. Three fundamental questions may be addressed. First, what are the overall policy objectives governing the management of uranium resources on federal lands? Second, is there a need for new legislation to modify or interpret these objectives? And third, what are some of the proposals which have been made?

Overall Policy Objectives

The Department of the Interior presently administers mineral leasing programs under four major statutes: the Mineral Leasing Act of 1920, the Acquired Lands Leasing Act of 1947, the Outer Continental Shelf Lands Act of 1953, and the Geothermal Steam Act of 1970.[113] The Department has generally considered its objectives under these acts to be threefold:

1. To ensure an orderly and timely development of the resource in question, in accordance with the Mining and Minerals Policy Act of 1970

2. To protect the environment, in accordance with the National Environmental Policy Act of 1969
3. To ensure the public a fair market value return on the disposition of its resources, in accordance with 31 U.S.C., Sec. 483.[114]

There can be conflicts among these objectives, and often it is necessary to balance various tradeoffs in order to arrive at a specific leasing policy decision.

The Federal Land Policy and Management Act of 1976 begins with a declaration of policy containing thirteen items. Among them are three which essentially restate the Interior Department's objectives, but limit the scope of the fair-market-value criterion:

> The Congress declares that it is the policy of the United States that . . .
> (8) the public lands be managed in a manner that will protect the quality
> of scientific, scenic, historical, ecological, environmental, air and atmo-
> spheric, water resource, and archeological values; that, where appro-
> priate, will preserve and protect certain public lands in their natural
> condition; that will provide food and habitat for fish and wildlife and
> domestic animals; and that will provide for outdoor recreation and
> human occupancy and use; (9) the United States receive fair market
> value of the use of the public lands and their resources unless otherwise
> provided for by statute; . . . (12) the public lands be managed in a manner
> which recognizes the Nation's need for domestic sources of minerals,
> food, timber, and fiber from the public lands including implementation of
> the Mining and Minerals Policy Act of 1970 [84 Stat. 1876, 30 U.S.C. 21a]
> as it pertains to the public lands. . . .[115]

This legislation is primarily directed toward the Bureau of Land Management and has been popularly known as the BLM Organic Act.

Although there are many other agencies involved in uranium supply, it is appropriate for us to limit our attention to the Department of the Interior in order to consider policies governing the disposition of resources on federal lands.

Need for New Legislation

It is obvious that under the Mining Law of 1872 the United States is not receiving "fair market value of the use of the public lands and their resources." However, the phrase "unless otherwise provided for by statute" is a loophole large enough to nullify the preceding phrase in the Federal Land Policy and Management Act.[116] We believe that the loophole should be removed by new legislation.

Section 102(b) of the Federal Land Policy and Management Act states that its policies shall "be construed as supplemental to and not in derogation of the purposes for which public lands are administered under other provisions of law.'' Since the purpose of the Mining Law of 1872 was basically to give federal sanctions to ''the local customs or rules of miners in the several mining districts,'' this sentence also may have been introduced in order to save the Mining Law of 1872 from being threatened by litigation.

Ironically, there exists a strong industry lobby (led by the American Mining Congress) pressing for orderly and timely development, and a strong environmental lobby pressing for reclamation and pollution controls, but no significant lobby concerned about receipt of fair market value.

Certainly the mining industry has reason for concern about timely development. There are already numerous state and federal environmental regulations which may involve delays and litigation and expensive alterations in plant and equipment. In addition, the radiation hazards associated with uranium mill tailings may involve additional health and safety standards. However, two points must be borne in mind. First, exploration expenditures have more than tripled in 3 years, going from $49.47 million in 1973 to $156.90 million in 1976;[117] and capital investment expenditures have more than doubled in 1 year, going from $254 million in 1976 to $535 million in 1977,[118] despite the fact that the rest of the nuclear industry has been practically stagnant since October 1974.[119] These data suggest that since 1973, the price increases in uranium concentrate have more than offset any increases in environmental costs and associated risks faced by the mining and milling industry. Second, it is doubtful that environmental regulations can be evaded, overridden, or weakened by some kind of ''timely development'' mandate. The only way to deal with these issues is through environmental legislation, litigation, and negotiation.

It is also clear that the public has reason to be concerned about environmental quality. We have already pointed out areas of difficulty in federal regulation. However, these concerns must be kept in perspective by recognizing three fundamental observations. First, many of the so-called environmental issues which are cited in lawsuits over implementation of the National Environmental Policy Act of 1969 are in fact health and safety issues covered by the Atomic Energy Act of 1954. They are the responsibility, not of the Interior Department, but of the Nuclear Regulatory Commission. Statutory authority is more than sufficient; the question is how the statutes should be implemented.[120] Second, a considerable amount of regulation exists at the state level, either through delegation of federal authority or through independent state authority. For example, the Union Pacific Railroad's mining and milling project required nine different permits or licenses from the Wyoming Department of Environmental Quality.[121] Third, the mandate of the National Environmental Policy Act is very broad, and many kinds of deficiencies in current regulatory practice can be corrected through the implementation of NEPA rather than the passage of additional legislation.

With regard to the fair-market-value objective, there is a clear need for new legislation. There are no considerations which diminish this need, for there is no way in which the government can capture the economic rents associated with its locatable mineral deposits unless the Mining Law of 1872 is changed or repealed.

In the case of uranium, the urgency of legislative reform has waxed and waned over years. Although there has been a strong case for reform since 1972, during the previous 40 years this was not so. It is instructive to distinguish six periods in the history of uranium mining in the United States:

1. From 1898, when radium was discovered by Pierre and Marie Curie, until 1930, when uranium deposits were found at Great Bear Lake, Canada, there was a strong demand for U.S. uranium ore. Radium is a product of the radioactive decay of uranium, and it is necessary to mine uranium in order to extract radium. Sandstone deposits in Colorado and Utah were found to be suitable. Although high-grade uranium deposits were found in Katanga, in the Belgian Congo, in 1913, the United States was the leading producer of radium from 1911 to 1923.[122] "These Colorado occurrences at one time provided annually several thousand tons of ore and, before the working of the Katanga ores, the United States produced more than twice as much radium as all other sources together."[123] Presumably, substantial economic rents were involved, but uranium and radium escaped the reach of the Mineral Leasing Act of 1920.

2. From 1930 until June 1940, when the National Defense Research Committee began taking steps to acquire uranium for military purposes,[124] the demand for U.S. uranium was practically eliminated by competition with low-cost foreign ores. It is interesting to note that an international cartel was formed at this time.

Before the war an agreement had been concluded between Union Miniere du Haut Katanga [operating in the Belgian Congo] and Eldorado Gold Mines Ltd. [Canada] for marketing radium, whereby the former would supply 60 per cent and the latter 40 per cent of the world requirements. This agreement was later dissolved. The production of uranium in 1939, principally from the Katanga and Great Bear Lake deposits, has been estimated at 1000 tons.[125]

This placed the United States in an awkward situation militarily in 1940:

There were no significant stockpiles in the United States, for the only commercial use of uranium was as a coloring agent in the ceramic industry. Of the 168 tons of oxides and salts American users consumed in

1938, only 26 came from domestic carnotite ores mined in the Colorado Plateau. The remainder was imported: 106 tons from the Belgian Congo and 36 from Canada.[126]

3. From 1940 until May 1958, when limited commercial sales were authorized, the federal government was the sole purchaser of domestically produced uranium concentrates. No other purchases were permitted, in order to ensure that supplies were available for military use. By this means it became possible for the government to capture all economic rent from its ore deposits. Although the AEC did not choose to do this, and thereby permitted great fortunes to be made in uranium, there was no lack of statutory authority to capture rents.

4. From May 1958 until 1966, when domestic uranium was first purchased for fueling commercial power reactors, the government purchased almost all our domestically produced uranium. On May 8, 1958 the AEC announced that it would permit U.S. companies to sell uranium concentrates to foreign and domestic buyers, although at that time it could not (under the Atomic Energy Act) permit private ownership of nuclear fuel. Shortly thereafter, a contract for export to the United Kingdom was announced.[127] Because of the availability of low-cost foreign uranium, the AEC's announcement had little impact on domestic production. However, the question of economic rents should have been addressed as soon as commercial sales were permitted.

5. From 1966 until February 1972, when a foreign uranium cartel was formed, domestic uranium was being purchased for use in commercial power reactors, but prices were so low that rents must have been very modest by today's standards. The Private Ownership of Special Nuclear Materials Act of 1964 permitted these purchases to begin. If receipt of fair market value had been a part of federal land policy at this time, the act should have included some provisions for uranium rents. However, the Mining Law of 1872 was of little concern to the Joint Committee on Atomic Energy.[128]

6. Since 1972 there has been an urgent need for reform of the location-patent system with respect to uranium. The reason is that an international cartel met in Paris in February 1972 in order to fix prices and allocate market shares. Several meetings were held subsequently. The U.S. Bureau of Mines described the meetings as follows.

Foreign countries with major uranium resources—Australia, Canada, France, and Republic of South Africa—sent representatives to Paris in February to consider the problems of oversupply and depressed prices for uranium in world markets. At a second meeting in March, also in Paris, the group called for rationalization of production or other marketing arrangements. At a third meeting, in Johannesburg, South Africa, in

June, an attempt was made to present a united front to uranium buyers, and a quota system and minimum price standards were under consideration.[129]

The impact on U.S. producers was immediately recognized by the Atomic Industrial Forum in the February 1972 issue of its journal:

At first sight, any arrangement that would put a floor under overseas prices looks like good news for U.S. producers. For one thing, higher prices abroad could open the way for new export sales. This possibility—that U.S. producers, now hardly a factor in the foreign market, might profit as much as anyone from any price-setting arrangement—is presumably one of the thornier problems that the four countries will try to find a way of solving.[130]

In fact, "a U.S. administration spokesman in 1972 reportedly favored price increases for uranium produced outside the United States to the extent they would improve the prospects of U.S. exports."[131]

It can be argued that U.S. producers have benefited from the cartel, not only in export sales, but in domestic sales. The details are complicated, however, and need not be discussed here. Our point is simply that the potential for obtaining economic rents from uranium claimed under the Mining Law of 1872 rose significantly in 1972, and the federal government was aware of the events underlying this trend.

Specific Proposals

In recent years there have been four significant proposals to repeal the Mining Law of 1872 and introduce a leasing system for hardrock minerals.[132] First, the proposed Mineral Leasing Act of 1973 was submitted to Congress by the administration on February 27, 1973 and was introduced as S. 1040 and H.R. 5442. The plan was to create a single leasing system for all minerals and to repeal the Mineral Leasing Act of 1920 as well as the Mining Law of 1872. The Department of the Interior and Department of Agriculture supported this proposal, but it was not approved by Congress.[133]

Second, the General Accounting Office issued a draft report early in 1974 on the need to reform the Mining Law of 1872. Comments were received from federal agencies in April 1974, but the report was not published until May 1975. The study called for a two-tiered system:

We recommend that the Congress enact legislation covering future exploration for and development of all minerals presently subject to the provisions of the Mining Law of 1872. This legislation would: Establish an exploration permit system covering public lands and require individuals interested in prospecting for minerals to obtain permits. Establish a leasing system for extracting minerals from public lands.[134]

Congress has not acted on the proposal.

Third, in October 1975 the Federal Trade Commission staff completed a report on energy resources on federal lands. The principal recommendation of the study was that two-stage competitive bidding be introduced in federal energy-leasing programs. Under such a system, unexplored tracts would be opened to competitive bidding for exploration rights, and tracts with known deposits would be opened to competitive bidding for development and production rights. One proposal regarding exploration rights was that these rights be awarded to the company or joint venture which will accept the lowest percentage share of whatever cash bonuses are paid for development and production leases on the tract of land to be explored. The FTC staff criticized the Mining Law of 1872 sharply, but did not outline a way to clear the government's title to old claims.[135]

Fourth, in March 1977 Representative Morris Udall, chairman of the Committee on Interior and Insular Affairs, introduced a bill to establish a two-tiered system for disposing of hardrock minerals similar to that proposed by the General Accounting Office. Prospecting permits would be issued for exploration, and leases for mining and milling. The Secretary of the Interior would establish detailed environmental regulations and set royalty rates. It is too soon to assess the future course of this bill.

In May 1977 the Council on Economic Priorities, a public-interest research organization, initiated a study of uranium resources on federal lands. The focus of the work is on leasing proposals, but no publications have yet been issued.

In summary, we cannot focus on a single proposal, as if it were the only alternative to the Mining Law of 1872, and analyze its implications. There are a variety of options before us. Accordingly, a reasonable objective for economic analysis in this context is to try to identify the different consequences of implementing different leasing systems, giving attention to a variety of alternatives. We shall attempt this task in chapter 9.

The Distribution of Uranium Lands and Federal Lands

As noted earlier, there is no public-domain land in Texas. This fact, together with the preceding data, suggest that the states with the greatest uranium potential are the five which cover most of the Wyoming Basins and all the Colorado Plateau—namely, Arizona, Colorado, New Mexico, Utah, and Wyoming. Since Alaska holds such a tremendous area of public-domain land, most of which has not been explored for uranium, we may add it to the list. These six states can be expected to play the leading roles in uranium production on federal lands.

Table 3–10 represents our basic guide to uranium lands and federal lands. The first six lines show how the area of the United States, and the areas of individual states, can be divided into four categories: public domain, acquired lands, patented lands with all minerals reserved to the United States, and other lands. These figures include certain lands held by the Bureau of Indian Affairs, which total 4,969,803 acres in the United States, but not all Indian lands, which total 53,175,374 acres in the United States.[136] The data on patented lands published by the Bureau of Land Management are not, unfortunately, reported on a state-by-state basis for the years following 1948. If they had been through fiscal year 1974, the figure in the fourth line of table 3–10, at the far right, would be 39,400,423 acres instead of 35,272,508 acres. In other words, only 89.5 percent of the 1974 total is represented in table 3–10. Moreover, 1,085,819 acres with "oil and gas plus other minerals" reserved (through 1974) and 2,306,584 acres with "miscellaneous minerals and combinations" reserved (through 1974) are not reported in table 3–10 because it is not clear how much, if any, of these contain a reservation of uranium.[137] Despite these omissions, the data in table 3–10 show that the federal government owns all mineral rights in at least 50.4 percent of the five-state area (Arizona, Colorado, New Mexico, Utah, and Wyoming).

The second part of table 3–10 describes ERDA's estimate of land held for uranium exploration, mining, and milling as of July 1, 1975. The five-state area holds 92.4 percent of the uranium mining claims, 95.1 percent of the acquired lands held for uranium, and 95.3 percent of the Indian lands held for uranium within the fourteen-state area covered by ERDA surveys. *Fee* land is held in "fee simple ownership," legally, and is just privately owned land. The third part of table 3–10 describes public land under exclusive jurisdiction of the Bureau of Land Management. All the vacant portion is open to uranium mining claims. All unreserved lands in Alaska were withdrawn under Public Land Order No. 5418 on March 25, 1974 so that they may be classified.[138] It is not clear what portion will be open to uranium mining, but a large area—perhaps half of Alaska—is likely to be involved.

The third line of table 3–10 describes total federal land in surface as well as subsurface ownership. This is reproduced as the bottom line of table 3–11, in which the holdings of various agencies are shown. The Bureau of Land Management is clearly the major landholder.

Table 3–12 shows the variation over time in the distribution of land

Table 3-10
Selected Data on Land Ownership and Use
(*in thousands of acres*)

	Arizona	Colorado	New Mexico	Utah	Wyoming	Total (5 States)[a]	Alaska	Other U.S.[b]	Total
Public domain (6/30/73)	31,628	22,875	24,259	34,322	29,442	142,526	353,367	208,859	704,751
Acquired lands (6/30/73)	306	1,064	1,608	543	626	4,148	17	52,083	56,248
Total federal land (6/30/73)	31,934	23,939	25,867	34,866	30,068	146,674	353,383	260,942	760,999
Patented lands with all minerals reserved (through 1948)	2,548	4,271	6,378	856	9,541	23,594	7	11,672	35,273
Other lands	38,206	38,275	45,521	16,975	22,734	161,712	12,092	1,301,258	1,475,072
Total area of state	72,688	66,486	77,766	52,697	62,343	331,980	365,482	1,573,882	2,271,343
Lands held for uranium (7/1/75):									
Claim	186	1,051	1,446	3,390	5,251	11,324		930	
Acquired	0	76	40	0	147	264		14	
Indian	0	0	425	90	180	604		30	
Total federal	186	1,127	1,911	3,390	5,578	12,192		973	
State	0	337	416	518	1,637	2,908		170	
Fee	700	171	1,322	100	2,528	4,821		1,213	
Total	886	1,652	3,649	4,008	9,744	19,938		2,357	
Public lands under exclusive jurisdiction of BLM (1974):									
Vacant, in grazing districts	9,935	5,911	11,073	20,130	10,928	57,977	0	74,333	132,310
Vacant, not in grazing districts	1,661	529	1,328	691	3,193	7,402	0	19,374	26,776
Reserved land	1,004	1,915	556	1,843	3,380	8,698	267,660	6,799	283,157
Unperfected entries pending	0	0	1	0	1	2	5,034	60	5,096
Total	12,601	8,355	12,958	22,664	17,502	74,079	272,694	100,566	447,339

Source: Land ownership data are from U.S. Department of the Interior, Bureau of Land Management, *Public Land Statistics, 1974* (Washington: GPO, 1975), pp. 10, 31, 46. Data on uranium lands are from U.S. Energy Research and Development Administration, "Survey of Lands Held for Uranium Exploration, Development and Production in Fourteen Western States in the Six Month Period Ending June 30, 1975," GJO-109 (71-1) (Grand Junction, Colo.: USERDA, 1976).

[a]Arizona, Colorado, New Mexico, Utah, Wyoming.

[b]Data on uranium lands refer to California, Idaho, Montana, Nevada, N. Dakota, Oregon, S. Dakota, Texas, and Washington.

Table 3-11
Federal Land Ownership as of June 30, 1973
(in thousands of acres)

	Arizona	Colorado	New Mexico	Utah	Wyoming	Total (5 States)[a]	Alaska	Other U.S. (excl. Alaska)	United States
Dept. of the Interior	16,878	9,225	13,855	24,934	20,766	85,657	329,962	123,534	539,153
Bureau of Land Management	12,587	8,332	12,949	22,641	17,543	74,052	299,130	100,602	473,784
Fish and Wildlife	1,537	48	94	98	44	1,822	19,756	6,505	28,083
National Park Service	1,603	529	241	888	2,311	5,572	7,006	12,104	24,682
Bureau of Reclamation	1,060	316	198	1,305	865	3,744	0	3,857	7,602
Bureau of Indian Affairs	90	0	373	0	1	466	4,065	440	4,970
Other Interior	0	0	0	0	1	1	5	27	33
Dept. of Agriculture	11,435	14,376	9,417	8,051	9,275	52,554	20,717	114,358	187,628
Forest Service	11,435	14,362	9,227	8,051	9,275	52,349	20,717	114,158	187,224
Agricultural Research Service	0	15	190	0	0	205	0	197	402
Other Agriculture	0	0	0	0	0	0	0	3	3
Dept. of Defense	3,620	301	2,537	1,876	26	8,359	2,524	19,733	30,616
Army	1,001	179	2,356	866	9	4,412	2,259	4,354	11,025
Air Force	2,582	29	165	921	7	3,705	136	4,546	8,387
Corps of Engineers	34	36	15	0	0	85	53	7,504	7,642
Navy	2	56	0	89	9	157	76	3,329	3,562
Atomic Energy Comm.	0	31	42	3	0	76	0	2,026	2,103
Tenn. Valley Authority	0	0	0	0	0	0	0	916	916
Dept. of Transportation	1	1	0	2	1	4	120	48	171
National Aeronautics and Space Admin.	0	0	4	0	0	4	9	124	137
International B&W Comm.	0	0	11	0	0	11	0	111	123
Other	1	6	0	0	0	8	52	91	152
Total	31,934	23,939	25,867	34,866	30,068	146,674	353,383	260,942	760,999

Source: U.S. Department of the Interior, Bureau of Land Management, *Public Land Statistics, 1974* (Washington: GPO, 1975), pp. 14–30.
[a]Arizona, Colorado, New Mexico, Utah, and Wyoming.

Table 3–12
Land Held for Uranium Exploration, Mining, and Milling in Fourteen Western States
(in millions of acres)

Category	1/1/71	1/1/72	1/1/73	1/1/74	1/1/75	1/1/76	1/1/77	1/1/78
Claim	11.471	8.755	9.679	10.290	11.634	12.605	15.067	16.594
Acquired	.555	.363	.206	.145	.275	.277	.293	.271
Indian	.674	.469	.603	.646	.635	.627	.815	.785
Total federal	12.700	9.587	10.488	11.081	12.544	13.509	16.175	17.650
State	5.835	3.995	1.859	1.945	2.968	3.385	4.635	5.048
Fee	5.871	5.425	5.330	5.748	5.764	6.017	6.273	6.613
Total	24.406	19.007	17.677	18.774	21.276	22.911	27.083	29.311

Sources: U.S. Energy Research and Development Administration, *Statistical Data of the Uranium Industry*, GJO-100(78) (Grand Junction, Colo.: USERDA, 1978), p. 67; *Survey of Lands Held for Uranium Exploration, Development and Production in Fourteen Western States in the Six Month Period Ending June 30, 1975*, GJO-109(76-1) (Grand Junction, Colo.: USERDA, 1976), p. 4.

Note: The states are Arizona, California, Colorado, Idaho, Montana, Nevada, New Mexico, North Dakota, Oregon, South Dakota, Texas, Utah, Washington, and Wyoming.

held for uranium exploration, mining, and milling. The decline and rise of state land is noteworthy.

The basic conclusion which one can draw from tables 3–10 and 3–12 is that the federal government appears to be in control of at least half the uranium-bearing lands in this country.[139] Table 3–11 shows that the Bureau of Land Management controls only about half the federal lands in Arizona, Colorado, New Mexico, Utah, and Wyoming; the other half is likely to be subject to conflicts between uranium production and other forms of land use. In short, federal policies regarding uranium lands will have a major effect on fuel availability for the nuclear industry.

Notes

1. U.S. Department of the Interior, Bureau of Land Management, *Public Land Statistics, 1975* (Washington: GPO, 1976), p. 36. See also 43 C.F.R. 3000.0-5(h).

2. U.S. Department of the Interior, *Public Land Statistics, 1975*, p. 35. See also 43 C.F.R. 3000.0-5(i).

3. U.S. Department of the Interior, Public Land Statistics, 1975, p. 36.

4. U.S. Federal Trade Commission Staff, *Staff Report on Mineral Leasing on Indian Lands* (Washington: GPO, 1976), pp. 1–2.

5. Philip Shabecoff, "Self-Rule by Indians Backed in U.S. Report," *New York Times*, March 17, 1977.

6. 10 C.F.R. 3811.2-9.

7. U.S. Federal Trade Commission staff, *Staff Report on Mineral Leasing on Indian Lands*, pp. 11–12.

8. 42 U.S.C., Sec. 2097.

9. This phrase appears in 10 C.F.R. 760.1(a).

10. U.S. Department of the Interior, *Public Land Statistics, 1975*, pp. 35, 104.

11. Robert S. Palmer, "Problems Arising Out of Public Land Withdrawals of the Atomic Energy Commission," *Rocky Mountain Mineral Law Institute* 2 (1956):80.

12. See 21 F.R. 5259 (July 14, 1956). Revisions have been published at 35 F.R. 17271 (November 10, 1970); 38 F.R. 21644 (August 10, 1973); and 41 F.R. 56783 (December 30, 1976).

13. The regulations were first published at 22 F.R. 1327 (March 5, 1957); and last published at 10 C.F.R. 60.9 (January 1, 1975).

14. C.F.R. (1976), Title 10, p. 465.

15. The full text of the lease is reproduced in U.S. Nuclear Regulatory Commission, U.S. Department of the Interior, and U.S. Department of Agriculture, *Draft Environmental Statement Related to Operation of Bear Creek Project, Rocky Mountain Energy Company*, NUREG-0129 (Washington: USNRC, 1977), pp. C-1 to C-11. Statutory authority for mineral leasing on acquired lands held by the Department of Agriculture is described in 43 C.F.R. 3500-3(b)(2)(i).

16. Clayton J. Parr and Northcutt Ely, "Mining Law," in *SME Mining Engineering Handbook*, edited by Arthur B. Cummins and Ivan A. Given (New York: Society of Mining Engineers, 1973), p. 2-20. See also 43 C.F.R. 3500-3(b)(2)(ii). Acquired lands "set apart for military or naval purposes" are not leased by BLM. See 43 C.F.R. 3501.2-1.

17. 10 C.F.R. 760.1(a).

18. These lands are in the states of Washington, South Carolina, and Idaho, respectively. See U.S. Department of the Interior, *Public Land Statistics, 1975*, p. 15.

19. Ibid., p. 37.

20. Ibid., p. 88.

21. U.S. General Accounting Office, *Modernization of 1872 Mining Law Needed to Encourage Domestic Mineral Production, Protect the Environment, and Improve Public Land Management*, B-118678 (Washington: USGAO, May 1975), pp. 24–30.

22. T.H. Watkins and Charles S. Watson, Jr., *The Lands No One Knows: America and the Public Domain* (San Francisco: Sierra Club Books, 1975).

23. Parr and Ely, "Mining Law," p. 2-18. See also 43 C.F.R. 3100.0-3(a)(1) and 3500.0-3(a)(1), (3), (4).

24. 43 C.F.R. 3500.0-3(a) (2), (5); and U.S. Federal Trade Commission staff, *Report to the Federal Trade Commission on Federal Energy Land Policy* (Washington: GPO, 1976), p. 59.

25. Robert W. Swenson, "Legal Aspects of Mineral Resources Exploitation," in *History of Public Land Law Development*, by Paul W. Gates and Robert W. Swenson (Washington: GPO, 1968), p. 751. See also U.S. Federal Trade Commission staff, *Report to the Federal Trade Commission on Federal Energy Land Policy*, pp. 82–83.

26. 30 U.S.C. 521 et seq.

27. This term is used in 43 C.F.R., Part 3740, where "Leasing Act" refers to the "mineral leasing laws" as defined in Section II of the Multiple Mineral Development Act of 1954.

28. 30 Stat. 36, cited in 43 C.F.R. 3811.1.

29. U.S. Department of the Interior, *Public Land Statistics, 1975*, p. 36.

30. See table 3–1 and ibid., p. 14.

31. See, for example, Jerry L. Haggard, "Regulation of Mining Law Activities on Federal Lands," *Rocky Mountain Mineral Law Institute* 21 (1975):361–376; and Randy L. Parcel, "Federal, State and Local Regulation of Mining Exploration," *Rocky Mountain Mineral Law Institute* 22 (1976):406–424.

32. 43 C.F.R. 3814.1(a). Conflicts with surface use are discussed in R.B. Holbrook, "Legal Obstacles to Uranium Development," *Rocky Mountain Mineral Lab Institute* 1 (1955):332–333, and in Parcel, "Federal, State and Local Regulation of Mining Exploration," pp. 425–426.

33. Henry C. Black, *Black's Law Dictionary*, revised 4th ed. (St. Paul, Minn.: West, 1970), p. 332.

34. See also 43 C.F.R. 3811.2-9. The complete regulations for the sale of land under color-of-title acts are found in 43 C.F.R., Part 2540.

35. Since money is not involved in the exchange, the lands are "conveyed" rather than "sold."

36. See 43 C.F.R., Part 2230 and 43 C.F.R. 3811.2-9. There are no regulations specifically addressed to this type of mining claim.

37. See 43 C.F.R. 2211.1-1, 43 C.F.R., Part 2220, and 43 C.F.R. 3811.2-9.

38. 43 U.S.C. 1719.

39. 43 C.F.R. 3816.2.

40. Battelle–Pacific Northwest Laboratories, *Assessment of Uranium and Thorium Resources in the United States and the Effect of Policy Alternatives*, PB-238 658 (Springfield, Va.: NTIS, 1974), p. 5.14.

41. 10 C.F.R. 40.4(h).

42. Ralph L. Miller and James R. Gill, "Uranium from Coal," *Scientific American* 191, 4 (October 1954):36–39.

43. 69 Stat. 679, Sec. 1.

44. 43 C.F.R. 3724.5(a).

45. 43 C.F.R. 3823.2(a).

46. Haggard, "Regulation of Mining Law Activities on Federal Lands," p. 372.

47. Swenson, "Legal Aspects of Mineral Resources Exploration," p. 750.

48. Cumulative totals through fiscal year 1974, based on the same data, are published in U.S. Federal Trade Commission staff, *Staff Report on Mineral Leasing on Indian Lands*, p. 8. Production on public lands covers only the 1963–1970 period.

49. 43 U.S.C. 1744.

50. James K. Groves, "Uranium Revisited," *Rocky Mountain Mineral Law Institute* 13 (1967):89–90.

51. "No lode located after May 10, 1872 can exceed a parallelogram 1,500 feet in length by 600 feet in width," according to 43 C.F.R. 3841.4-2. A rectangle with these borders has a maximum possible area.

52. 43 C.F.R. 3842.1-2.

53. USERDA, *Statistical Data of the Uranium Industry*, GJO-100 (77) (Grand Junction, Colo.: USERDA, 1977), p. 86.

54. U.S. Federal Trade Commission staff, *Report to the Federal Trade Commission on Federal Energy Land Policy*, p. 674.

55. The act of March 20, 1922, as amended, provides statutory authority. See 43 C.F.R., Part 2230.

56. The Federal Land Policy and Management Act of 1976 provides statutory authority. See 43 U.S.C. 1719.

57. Richard A. Clark, Robert C. Lind, and Robert Smiley, *Enhancing Competition for Federal Coal Leases*, SAI-76-513-WA (McLean, Va.: Science Application, Inc., 1976), pp. 56–85.

58. A more complete discussion can be found in Swenson, "Legal Aspects of Mineral Resources Exploitation." This is the basic reference for this historical sketch.

59. U.S. Department of the Interior, *Public Land Statistics, 1975*, p. 4.

60. Swenson, "Legal Aspects of Mineral Resources Exploitation," p. 701.

61. Ibid., p. 707.

62. 43 U.S.C. 1701.

63. Swenson, "Legal Aspects of Mineral Resources Exploitation," p. 745.

64. U.S. Federal Trade Commission staff, *Report to the Federal Trade Commission on Federal Energy Land Policy*, p. 4.

65. *Congressional Record* 58 (1919):4111, quoted in Swenson, "Legal Aspects of Mineral Resources Exploitation," p. 742.

66. U.S. Federal Trade Commission staff, *Report to the Federal Trade Commission on Federal Energy Land Policy*, p. 55. The quotations are from *Congressional Record* 58 (1919):4253.

67. Ibid., p. 69.

68. See U.S. Department of the Interior, *Public Land Statistics 1975*, p. 10. Acquired lands are included in these figures.

69. Parcel, "Federal, State and Local Regulation of Mining Exploration," pp. 424–425.

70. 43 C.F.R. 3501.2-5(b)(2)(ii).

71. 22 F.R. 1327 (March 5, 1957); 43 C.F.R. 3101.1-5 and 3101.2-4.

72. U.S. Federal Trade Commission staff, *Staff Report on Mineral Leasing on Indian Lands*, pp. 59–60.

73. Implicit price deflators for GNP are found in U.S. Council of Economic Advisors, *Economic Report of the President, 1975* (Washington: GPO, 1975), p. 253.

74. "Uravan Pre-Bid Meeting," *Nuclear Industry*, January 1974, p. 40.

75. 10 C.F.R. 60.8 (1956), first published in 21 F.R. 5259 (July 14, 1956).

76. "AEC to Consider Making Its Uranium Reserve Available for Leasing," *Nuclear Industry*, July 1970, pp. 29–31.

77. 10 C.F.R. 760.1 (1977), first published in 38 F.R. 21644 (August 10, 1973).

78. "Uravan Pre-Bid Meeting," p. 40.

79. U.S. Federal Trade Commission staff, *Staff Report on Mineral Leasing on Indian Lands*, pp. 45, 60, 64.

80. "AEC to Consider Making Its Uranium Reserve Available for Leasing," p. 31.

81. 10 C.F.R. 60.5a(a) (2), as published in 1975.

82. Swenson, "Legal Aspects of Mineral Resources Exploitation," p. 709.

83. Gregory Yale, *Mining Claims and Water Rights* (1867), pp. 332–336, quoted in Swenson, "Legal Aspects of Mineral Resources Exploitation," p. 719.

84. Swenson, "Legal Aspects of Mineral Resources Exploitation," p. 751.

85. Groves, "Uranium Revisited," p. 112.

86. Holbrook, "Legal Obstacles to Uranium Development," p. 333.

87. 30 U.S.C. 22 (1970).

88. Swenson, "Legal Aspects of Mineral Resources Exploitation," p. 723.

89. Holbrook, "Legal Obstacles to Uranium Development," p. 335.

90. *Book* v. *Justice M. Co.*, 58 Fed. 106 (CCD Nev. 1893), quoted in Holbrook, "Legal Obstacles to Uranium Development," p. 336.

91. *Castle* v. *Womble*, 19 L.D. 455 (1894), quoted in Holbrook, "Legal Obstacles to Uranium Development," p. 339.

92. U.S. Federal Trade Commission staff, *Report of the Federal Trade Commission on Federal Energy Land Policy*, p. 666.

93. William G. Waldeck, "Legal Problems Affecting Uranium Mining," in *Uranium and the Atomic Industry* (New York: Atomic Industrial Forum, Inc., 1956), p. 179.

94. Groves, "Uranium Revisited," p. 94.

95. 10 C.F.R. 60.5a (b) (1), as published in 1975.

96. Swenson, "Legal Aspects of Mineral Resources Exploitation," p. 732.

97. Holbrook, "Legal Obstacles to Uranium Development," p. 349.

98. U.S. General Accounting Office, *Modernization of 1872 Mining Law Needed to Encourage Domestic Mineral Production, Protect the Environment, and Improve Public Land Management*, p. 45.

99. William S. Greever, *The Bonanza West: The Story of the Western Mining Rushes, 1848–1900* (Norman, Okla., 1963), p. 107, quoted in Swenson, "Legal Aspects of Mineral Resources Exploitation," p. 719; see also pp. 710–711.

100. U.S. General Accounting Office, *Modernization of 1872 Mining Law*, p. 33.

101. Ibid., p. 14.

102. Swenson, "Legal Aspects of Mineral Resource Exploration," p. 736.

103. Address by F.C. Love, National Western Mining Conference, February 9, 1967, cited in Groves, "Uranium Revisited," p. 87.

104. USERDA, *Statistical Data of the Uranium Industry*, GJO-100(77), pp. 21–22. The first figure refers to reserves which can be extracted at a "forward cost" no greater than $50 per pound of U_3O_8 in 1977 dollars. Forward costs will be discussed in chapter 5.

105. U.S. Congress, Joint Committee on Atomic Energy, *Atomic Energy Legislation through 94th Congress, 2d Session* (Washington: GPO, 1977), p. 362.

106. Ibid., p. 32.

107. Ibid., pp. 113, 123. The transfers are made in Sections 201(f) and 104(c), respectively.

108. Groves, "Uranium Revisited," p. 98.

109. 10 C.F.R. 51.5(a) (5), 51.5(b) (5).

110. 42 U.S.C., Sec. 2021.

111. Information about this case has been provided by the Natural Resources Defense Council in private correspondence.

112. 10 C.F.R. 51.1(a). The wording is based on NEPA.

113. 30 U.S.C., Secs. 181–287; 30 U.S.C., Secs. 351–359; 43 U.S.C., Secs. 133–1343; and 30 U.S.C., Secs. 1001–1025, respectively.

114. See Robert J. Kalter and Wallace E. Tyner, "Disposal Policy for Energy Resources in the Public Domain," in *Energy Supply and Government Policy*, edited by Robert J. Kalter and William A. Vogely (Ithaca, N.Y.: Cornell Univ. Press, 1976), pp. 52–53; and U.S. Federal Trade Commission staff, *Report on the Federal Trade Commission on Federal Energy Land Policy*, p. 211.

115. 90 Stat. 2744, 43 U.S.C. 1701.

116. Ibid.

117. William L. Chenoweth, "Exploration Activities," *Uranium Industry Seminar, October 19–20, 1976,* GJO-108 (76) (Grand Junction, Colo.: USERDA, 1976), p. 180.

118. "Uranium Producers in U.S. Doubling Capital Outlay," *ERDA News* 2, 12 (June 13, 1977):1. Exploration expenditures are not included in these figures.

119. We will discuss nuclear plant capacity growth in chapter 4.

120. We disagree with the NRC's policy that uranium mines should not be licensed because they involve only "unimportant" quantities of source material.

121. U.S. Nuclear Regulatory Commission, U.S. Department of the Interior, and U.S. Department of Agriculture, *Final Environmental Statement Related to Operation of Bear Creek Project, Rocky Mountain Energy Company*, NUREG-0129 (Washington: USNRC, 1977), p. 1-6.

122. Walter C. Woodmansee, "Uranium," in *Mineral Facts and Problems*, issued by U.S. Department of the Interior, Bureau of Mines (Washington: GPO, 1976), p. 1177.

123. C.E. Tilley, "Raw Materials for Atomic Power," in *Atomic Energy*, edited by J.L. Crammer and R.E. Peierls (Harmondsworth, England: Penguin, 1950), p. 99.

124. Richard G. Hewlett and Oscar E. Anderson, Jr., *The New World* (Washington: USAEC, 1972), p. 26.

125. Tilley, "Raw Materials for Atomic Power," p. 98.

126. Hewlett and Anderson, *The New World*, p. 26.

127. U.S. Department of the Interior, Bureau of Mines, *Minerals Yearbook, 1958* (Washington: GPO, 1960).

128. U.S. Congress, Joint Committee on Atomic Energy, *Private Ownership of Special Nuclear Materials*, Hearings, 88th Cong., 1st Sess., July 30, 31, and August 1, 1963 (Washington: GPO, 1964); and *Private*

Ownership of Special Nuclear Material, Hearings, 88th Cong., 2d Sess., June 9, 10, 11, 15, and 25, 1964 (Washington: GPO, 1964).

129. Walter C. Woodmansee, "Uranium," in *Mineral Yearbook, 1972*, issued by U.S. Department of the Interior, Bureau of Mines (Washington: GPO, 1974), pp. 1272–1273. This report was based on the article, "Producers Meet in Johannesburg," *Mining Journal* 278, 7140 (June 23, 1972): 514–515.

130. "Big Four Uranium Producers Weigh 'Orderly Marketing' Pact," *Nuclear Industry*, February 1972, p. 42.

31. Zuhayr Mikdashi, *The International Politics of Natural Resources* (Ithaca, N.Y.: Cornell Univ. Press, 1976), p. 109. This statement is based on reports in the *Australian Financial Review*, August 25, 1972, p. 1, and August 28, 1972, p. 10.

132. This term must be used in the context of reform, since it would be confusing to refer to locatable minerals in proposals to abolish the location system.

133. U.S. General Accounting Office, *Modernization of 1872 Mining Law*, pp. 4, 46.

134. Ibid., p. 44.

135. U.S. Federal Trade Commission staff, *Report to the Federal Trade Commission on Federal Energy Land Policy*, pp. 733–736, 760–763.

136. U.S. Federal Trade Commission staff, *Staff Report on Mineral Leasing on Indian Lands* (Washington: GPO, 1976), p. 5.

137. U.S. Department of the Interior, Bureau of Land Management, *Public Land Statistics, 1974* (Washington: GPO, 1975), p. 46.

138. Ibid., p. 31.

139. This point is confirmed by data referenced in note 9, chapter 9.

4 Domestic Uranium Demand

We will approach the subject of domestic uranium demand by distinguishing four topics. First, the various factors which influence uranium demand will be discussed, with the reasons why we prefer to use projections based on the assumption that demand is inelastic with respect to uranium prices. Second, the fuel requirements of individual reactors will be examined on the basis of engineering data. Next, the way in which a utility purchases uranium to cover projected fuel requirements will be examined. The final ingredient in a demand projection is, of course, the schedule of nuclear capacity additions. We choose to address these topics in this order because estimates of fuel requirements are based on relatively "hard" data, while nuclear capacity projections are highly subjective and uncertain. From this perspective, we will briefly examine official projections of uranium demand.

The Nature of Uranium Demand

Every step in the nuclear fuel production process involves a market for which supply and demand can be evaluated. However, the market for uranium concentrates (U_3O_8) is generally given far more attention than any other. Since revenues from federal leasing may be based on ore values, we shall look at the uranium ore market in chapter 7. Since the demand for concentrates is derived from the demand for uranium hexafluoride (UF_6), we shall also examine the latter and the way it is influenced by federal policy decisions.

Components of Uranium Concentrate Production

First, let us consider U_3O_8. The total quantity produced in a given year can be divided into three components. One portion is needed to fuel the reactor capacity which will be in operation 6 to 18 months after the year in question, assuming a 1.5-year lead time, on average, between the production of uranium in U_3O_8 and its loading into a reactor. The Atomic Energy Commission estimated a 1-year lead time between U_3O_8 procurement and charging of reload fuel and a 1.75-year lead time between U_3O_8 procure-

ment and commercial operation with initial cores.[1] A more recent estimate yields a 1.25-year lead time between U_3O_8 procurement and charging
of reload fuel and a 1.75-year lead time between U_3O_8 procurement and
loading of the initial core.[2] Since the actual operating capacity is known
only in retrospect, a precise measurement of this portion of U_3O_8 requirements can be calculated only retrospectively.

Another portion must be delivered in a given year to fulfill contract
commitments in excess of the requirements based on actual operating
capacity. These commitments may be based on optimistic projections of
nuclear power plant construction and licensing schedules, or they may be
related to plans for stockpiling uranium in some form other than ores and
ore reserves.

A third portion consists of additions to the stockpiles of U_3O_8 held by
producers. These additions do not include inventories held in a given year
for deliveries the same year under contract commitments. Producers are
likely to build up these stockpiles whenever the net present value of
after-tax profits is maximized under a stockpiling plan as opposed to a
schedule of production just prior to delivery. The difference in net present
values is likely to be caused by differences in mining costs, although
royalties, taxes, and other factors may be important.

*Importance of Utility Commitments to Reactor
Construction*

The first two of these components may be said to comprise U_3O_8 demand.
This demand is much less elastic with respect to ''committed'' nuclear
capacity—that is, that for which financial commitments have already been
made—than ''uncommitted'' capacity. According to a representative of a
West German utility company, uranium demand involving committed
reactors is perfectly inelastic:

> Once a nuclear power plant has been ordered, is under construction or is
> actually in operation . . . the uranium necessary for the plant's operation
> must be obtained at all costs. A uranium price so high as to require the
> shutdown of a nuclear power plant is inconceivable.[3]

We are unaware of any study which estimates ''a uranium price so high as
to require the shutdown of a nuclear power plant.'' Moreover, we are
unaware of any estimate of a uranium price so high as to require that an
investment in nuclear plan design and construction be written off as a
loss.

Since the scale of a utility's commitment to a nuclear plant varies
along a continuum, the elasticity of uranium demand related to the plant

also varies. Moreover, different utilities will have different subjective assessments of long-term price and cost trends and different risk preference functions. In the case of reactors for which a construction permit application has not been filed, experience shows that demand is clearly elastic. Three examples demonstrate this point. In September 1975, South Carolina Electric and Gas announced cancellation of V.C. Summer, Unit 2 and stated that one reason for the decision was that an "acceptable" supplier of uranium ore could not be found.[4] In December 1975, Florida Power Corporation canceled its letter of intent to order two reactors and stated that uncertainties regarding uranium availability were a major concern.[5] In January 1976 the Sacramento Municipal Utility District announced cancellation of Rancho Seco 2, citing three reasons, the second of which involved uranium prices: "Two years ago we estimated the price of yellowcake at $17 a pound [this was in 1984, with 1984 dollars]; today it cannot be purchased for less than $30 a pound and it is not easy to find suppliers even at that price."[6] None of these reactors were undergoing Nuclear Regulatory Commission review.[7]

A nuclear plant may be cancelled even after a large financial commitment has been made by the utility involved. For example, Virginia Electric and Power Company abandoned Surry 3 and 4 after committing $146 million:

> The company said the decision to cancel the units was made "in light of lower demand forecasts for future years, growing concern over the many uncertainties that face the nuclear industry at this time, together with the increasing financial burdens these uncertainties impose on utilities and their customers."[8]

Among these uncertainties, uranium supply was surely a major concern, since Virginia Electric had already sued Westinghouse Electric Corporation for fulfillment of uranium supply contracts. The utility asked that the court either order fulfillment or award damages of $250 million.[9]

Whenever cancellation of a reactor can be seriously considered, the nuclear plant, the uranium concentrates to fuel it, the nuclear fuel-cycle services needed to process these concentrates, and operation and maintenance services are all complementary goods and services whose markets are interrelated. An upward shift in the supply function for any one of these will tend to reduce the cost advantage of nuclear plants and thereby reduce demand for all the other goods and services. When cancellation is out of the question, U_3O_8 demand is still related to the markets for fuel-cycle services and the cost of operating and maintaining the reactor. If the utility owns power plant capacity which can substitute for a particular nuclear plant, this other capacity will be used when it can be fueled, operated, and maintained more cheaply than the nuclear plant.

Today, such a situation is unknown in the utility industry, although it may develop as nuclear plants grow older.

Uranium demand is further complicated by the fact that uranium is generally ordered many years in advance of delivery. Different shipments delivered in a given year may have very different prices, because of different ordering dates. Demand elasticity decreases as the lead time before delivery decreases. In the spot market, demand is highly inelastic, since purchases often involve uranium which must be processed and used immediately.

Modeling of Elastic Demand

The price of uranium affects not only the way uranium is purchased and stockpiled, but also the demand for nuclear reactors, the demand for base-load capacity generally, the demand for electricity, and ultimately, the demand for energy. In theory it would be possible to model the uranium market by balancing supply and demand in a complex energy model. A demand curve could be generated by running the model several times, using different input assumptions about the uranium supply curve. An iterative procedure based on the following steps might be employed:

1. Estimate future annual energy use.
2. Estimate the share of this energy which will be electrical, given the relative prices of electricity and other energy sources.
3. Estimate the schedule of base-load generating additions required to meet this electricity demand projection.
4. Estimate the nuclear share of these capacity additions, given capital costs, uranium prices, coal prices, and so forth.
5. Estimate uranium requirements for these capacity additions.
6. Estimate the additional uranium production which will be attributable to contract coverage in excess of actual requirements and to stockpiling.

The MIT Regional Electricity Model covers the first five steps, but it assumes exogenous uranium prices (perfectly elastic uranium supply curves). According to Joskow and Baughman, a doubling of the prices of uranium concentrates and enrichment services caused a 12.8 percent reduction in cumulative 1974–1995 uranium utilization relative to the base case. No report was made on the effect of uranium price increases alone, nor were high uranium prices (above $30 per pound in 1974 dollars) evaluated. Although the uranium supply curves were taken to be perfectly elastic, they were assumed to shift upward as a function of time.[10]

Modeling of Inelastic Demand

An alternative method of modeling uranium demand is to assume that it is perfectly inelastic and to go through the six steps just listed. The results can then be used in conjunction with a model of elastic (but not necessarily perfectly elastic) uranium supply to generate uranium price projections. One such inelastic demand model was developed by the Energy Research and Development Administration and its contractors. Only the first five of the preceding steps are covered.[11]

At present it appears likely that nuclear fuel reprocessing will be indefinitely deferred in the United States. Under these conditions, the model used by ERDA projected an installed nuclear capacity of 195 GWe (gigawatts, electrical) by 1990.[12] ERDA reported that a total of 203 GWe was operable, under construction, or on order as of June 1977.[13] By contrast, the base case of the MIT model projects a capacity of 406 GWe by 1990, given nuclear plant lead times of 10 years.[14] This scenario is extremely unlikely, for it would require 203 GWe of new orders (net of cancellations) over the next 3 years and no increase in lead times above their current level.

We choose to use DOE's nuclear capacity projection (covering the first four of the preceding steps) as the basis for projections of uranium demand, assuming price inelasticity. There are several reasons why we will not attempt to model elastic demand. First, it is much simpler to use capacity projections from a model of inelastic demand than to use or develop a model of elastic demand. The focus of this research is on uranium supply, particularly that which involves federal lands, and reactor demand modeling is too large a task to be included within this scope. We could not use the MIT model in its present form because it involves the assumption that uranium supply is perfectly elastic.

Second, the price elasticity of uranium demand cannot be estimated by making extrapolations on the basis of past experience, that is, by fitting a regression model and assuming that no major changes in utility-ordering behavior will occur. There are three reasons for this: (1) the secrecy accompanying uranium price data associated with particular reactors, (2) the variability of other prices affecting nuclear reactor ordering, and (3) the rapid increase in uranium prices which occurred during 1974–1976. Consequently, a model designed to simulate future utility ordering cannot be adequately tested to demonstrate its validity.

Third, we have reason to believe that the price elasticity of uranium demand will be small between now and the year 2000 in the United States. The October 1978 "high" projection by DOE involves 395 GWe of installed nuclear capacity in the year 2000,[15] of which 199 GWe, or 50 percent, was operable, under construction, or on order.[16]

The remaining 196 GWe cannot begin operation until the 1990s and must, therefore, represent much less than 50 percent of cumulative 1977–2000 deliveries of U_3O_8. The cost of U_3O_8 is likely to represent only about 10 percent of the busbar cost of electricity from these nuclear plants. In other words, a 10 percent change in uranium prices would cause only a 1 percent change in nuclear power costs, which would have only a slight influence on new reactor orders. I estimate that U_3O_8 will cost roughly 3 mills/kWeh out of a total of 30 mills/kWeh in 1977 dollars.

Fourth, there are many factors other than uranium prices which can cause major shifts in uranium demand. The other 90 percent of busbar cost is subject to price increases, and the licensing procedures for nuclear facilities involve a great many political and technical questions which go beyond strictly economic calculations. Thus it would be imprudent to devote a major research effort to the slope of the uranium demand curve, when shifts in the curve can exert at least as great an influence on uranium prices.

And last, it is significant that despite all the sophistication and complexity of the MIT model, the base case yielded a capacity projection which now appears totally unreasonable. While it would no doubt be possible to change various parameters so that the level of 1977–1980 reactor orders would shrink to a more credible range, the forecasting accuracy of the model can still be seriously questioned. Recent experience suggests that we cannot place much confidence in the model's predictions because we cannot forecast the input parameters accurately, even for a particular "case" or scenario. This is a problem involved in any model of elastic demand. Moreover, these difficulties are even more severe in the case of foreign uranium demand. Thus Chapter 8 will deal only with projections based on inelastic uranium demand models.

Fuel Requirements of Individual Reactors

Cumulative reactor operating experience to date is not sufficient to predict future requirements with certainty, but it suggests that domestic reactors will generate less electricity per ton of U_3O_8 than they have been designed to produce.

Enrichment Tails Assay

After U_3O_8 has been converted to UF_6, it must be "enriched," that is, the uranium *feed*, or input to the enrichment plant, must be divided into two quantities, a *product*, in which the concentration of the isotope U-235 is higher than that found in nature, and *tails*, in which the U-235 concentration is lower than that found in nature. The natural concentration varies slightly, but a standard of 0.711 percent by weight is used by DOE.[17] Losses in the enrichment process are negligible, so the total amount of U-235 in feed equals that in the product and tails:

$$0.711f = ep + t(f - p) \qquad 0 < t < 0.711 \tag{4.1}$$

where f = quantity of uranium feed, in kg U
$\quad\quad p$ = quantity of enriched uranium, in kg U
$\quad\quad e$ = enrichment level of product, in weight percent U-235
$\quad\quad t$ = tails assay, in weight percent U-235
$\quad$ kg U = kilograms of uranium

This equation determines the ratio of feed to product:

$$\frac{f}{p} = \frac{e - t}{0.711 - t} \qquad \frac{e}{0.711} < \frac{f}{p} < \infty \tag{4.2}$$

The fraction of U-235 in the feed which ends up in the product is called the *U-235 utilization rate*.[18]

The value of e is determined by the utility customer or nuclear reactor manufacturer according to the design specifications of the particular reactor for which enriched uranium is being provided. In planning for the typical light-water reactor, 3.0 percent is an appropriate figure. The value of t can be manipulated by the federal government, which owns all the enrichment plants in the United States and is planning to build another one. To be precise, there are two values of t: the *transaction tails assay*, which is used for computing the quantity of feed that must be delivered to DOE in order to obtain a certain amount of product, and the *operating tails assay*, which is the actual value used in operating the enrichment plants. It follows that there are two kinds of uranium requirements: *transaction requirements*, which industry must provide to DOE, and *operating requirements*, which measure actual uranium concentrate use. It is possible that one or more privately owned enrichment plants will be operated in the future, perhaps by 1988.[19] Any discussion of the tails assay for these plants must be highly speculative, although in 1976 ERDA experts believed "that 0.20 percent U-235 is an appropriate figure for planning purposes."[20]

To develop uranium demand projections, it is necessary to know the transaction tails assay which will be used by DOE. This is no simple task. On November 29, 1967, the transaction tails assay was set at 0.20 percent U-235, effective January 1, 1969, when the government began providing enrichment services to private industry.[21] Since 1971 the government has issued several proposals for increasing the transaction tails assay, but no increase has ever been put into effect, although the operating tails assay was raised from 0.20 percent to 0.30 percent, effective July 1, 1971, and then lowered to 0.25 percent, effective July 1, 1975. To understand these various proposals and changes, we must briefly examine enrichment technology.

Enriched uranium is a product made from two substitutable inputs, UF_6 feed and separative work. The number of separative work units (SWU) per kilogram of product is called the SWU/product ratio and depends on the enrichment level and tails assay. The SWU/product ratio is inversely related to the feed/product ratio:[22]

$$\frac{s}{p} = \left(\frac{f}{p} - 1\right)\left(\frac{t}{50} - 1\right)\ln\frac{t}{100 - t} + e\left(\frac{e}{50} - 1\right)\ln\frac{e}{100 - e}$$

$$- \frac{f}{p}\left(\frac{0.711}{50} - 1\right)\ln\frac{0.711}{100 - 0.711}$$

$$= \left(\frac{f}{p} - 1\right)\left(\frac{t}{50} - 1\right)\ln\frac{t}{100 - t} \tag{4.3}$$

$$+ e\left(\frac{e}{50} - 1\right)\ln\frac{e}{100 - e} - 4.8689\frac{f}{p}$$

where s/p = SWU/product ratio, in SWU/kg U

From the utility industry's viewpoint, it is desirable to minimize the cost of purchasing enriched uranium:

$$C_{EU} = C_{UF_6}\left(\frac{f}{p}\right) + C_{SWU}\left(\frac{s}{p}\right) \tag{4.4}$$

where C_{EU} = cost of enriched uranium, in \$/kg U
C_{UF_6} = cost of UF_6 feed, in \$/kg U
C_{SWU} = cost of a separative work unit, in \$/SWU
s/p = SWU/product ratio, in SWU/kg U

From the mining industry's viewpoint, it is desirable to maximize the feed/product ratio. The selection of tails assays by the federal government is largely an effort to compromise between these two viewpoints.

When the cost of a separative work unit rises faster than the cost of UF_6 feed, the utility industry may desire to lower the SWU/product ratio. This can be accomplished by raising the tails assay. A rise in the tails assay will raise the feed/product ratio and thereby increase the demand for UF_6. This increased demand will please the mining industry. On the other hand, when separative work is cheap, utilities will want to pay for enrichment services at a low tails assay and miners will want the government to set a high tails assay. The historical trend in separative work costs for requirements-type contracts is shown in figure 4-1. From February 25, 1971 to April 24, 1979, costs increase at an annual rate of 16.3 percent. A high rate of increase also applies for fixed-commitment contracts.

During the years 1964–1970, the Atomic Energy Commission purchased 56,242 short tons of U_3O_8 from domestic producers, and in 1970 a stockpile surplus of about 50,000 short tons remained.[23] These purchases were made in order to subsidize the uranium mining industry during the "lean years" when military requirements were declining and civilian requirements had not yet grown to a substantial level. In 1971 the government held both a surplus of enrichment capacity and a surplus of U_3O_8. It decided to use up the surplus of U_3O_8 by setting the operating tails assay above the transaction tails assay. The amount of feed required to operate the enrichment plants was thereby raised above the amount delivered by utility companies. The price of separative work included an allowance for depreciation on the enrichment plants. Therefore, separative work was priced above its marginal cost, which was essentially just an electric power cost. Had it been priced at marginal cost, the utilities would probably have objected to the high transaction tails assay. In the future, the stockpile will run out, and then the government will have to set the transaction tails assay equal to the operating tails assay. Moreover, in the future there may be a shortage of enrichment capacity, relative to the amount needed to sustain a 0.20 percent operating tails assay. In such a situation, the operating tails assay must be increased. This is why the proposals to raise the transaction tails assay have been made.

As nuclear capacity projections have fallen, the government has postponed the date at which its U_3O_8 stockpile is projected to run out and the date at which a shortage of enrichment capacity is projected. Consequently, increases in the transaction tails assay have been postponed.

As of June 1977, ERDA held enrichment contracts for 208 GWe of domestic nuclear capacity plus 115 GWe of foreign nuclear capacity. The three existing diffusion plants, plus the new plant proposed by President Carter,[24] can supply the necessary amount of annual separative work for these reactors indefinitely at an operating tails assay of 0.25 percent U-235.[25] However, in July 1977 the acting administrator of ERDA said that the new plant will permit the government to sign enrichment contracts for additional reactor capacity.[26] Such contracts would eventually require a higher operating tails assay.

The utility industry would prefer to select the transaction tails assay on a short-term basis rather than having the government establish a

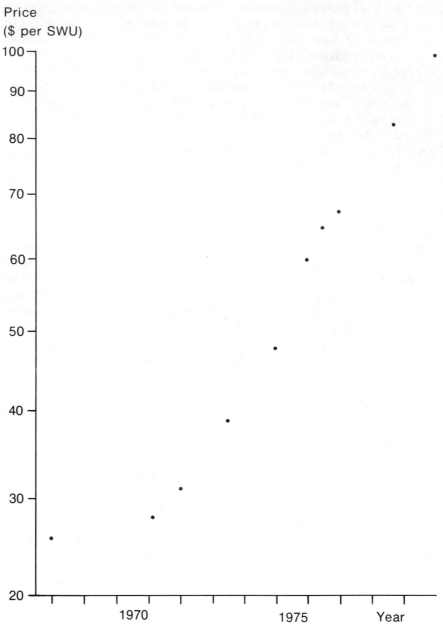

Figure 4–1. Charges for Enrichment Services under Requirements-Type
Contracts.

Sources: 32 F.R. 16289; 35 F.R. 13457; 36 F.R. 4563; 38 F.R. 4432; 38 F.R. 21518; 39 F.R.
22182; 40 F.R. 26061; 41 F.R. 8414; 41 F.R. 3942; 42 F.R. 51635; 43 F.R. 49831.

The data given here are based on prices published in *Federal Register* notices and do not
take "ceiling charges" into account.

long-term plan. This contracting flexibility would allow utilities to mini-mize the cost of enriched uranium in response to short-term fluctuations in the cost of UF_6 and separative work and to specify a low feed/product ratio whenever it becomes difficult to obtain large quantities of UF_6.[27] Price and quantity uncertainties regarding UF_6 are almost entirely based on uncertainties about U_3O_8. Price uncertainties regarding separative work are largely based on technological change, and on federal incentives for development of privately owned enrichment facilities.

There are many political, economic, and technical issues which create uncertainty about the future level of tails assays for both federally owned and privately owned enrichment plants. We choose not to discuss these issues further, since they do not add anything to the foregoing discussion which would enable us to project future tails assays with more confidence.

Duty Factor

Once a feed/product ratio has been established, the next step in calculat-ing uranium requirements is to estimate the duty factor, or number of net kilowatt-hours (electrical) generated per short ton (ST) of U_3O_8 in con-centrate (not in ore reserves). Let us trace the use of a particular short ton of U_3O_8 through the fuel cycle. The yellowcake is converted to UF_6 and then measured in terms of kilograms of uranium. Only a portion of the feed ends up as product. Some fraction of the uranium in U_3O_8 will be lost in the nuclear fuel production process. The remainder of the enrichment product will be "burnt" in a reactor, releasing heat. The amount of heat generated per unit of uranium is measured in kilowatt-days (thermal) per kilogram of uranium, and is called the *burnup*. Some fraction of this heat will be converted to electrical energy, which is measured in kilowatt-hours rather than kilowatt-days. Thus the duty factor can be calculated simply:

$$D = 770.89 \frac{\text{kg U}}{\text{ST U}_3\text{O}_8} \cdot \frac{p}{f} (1 - Z)BE \cdot 24 \, h/d$$

$$= \frac{0.711 - t}{e - t} (1 - Z)BE \cdot 1.850 \cdot 10^4 \frac{\text{kg U} \cdot h}{\text{ST U}_3\text{O}_8 \cdot d} \qquad (4.5)$$

where D = duty factor, in kWeh/ST (short ton) U_3O_8
     p/f = product/feed ratio; the inverse of the feed/product ratio
     Z = fraction of uranium in U_3O_8 which is lost in the nuclear fuel production process
     B = burnup, in kWtd/kg U

E = conversion efficiency, or net electrical power per unit of gross thermal power, in kWe/kWt

h = hour

d = day

kWe = kilowatt (electrical)

kWt = kilowatt (thermal)

For a typical reactor we can assume values of $E = 0.32$ kWe/kWt and $Z = 0.015$.[28] If we assume further that $e = 3.0$, the duty factor can be calculated as a function of the tails assay and burnup, as shown in figure 4–2. For convenience, the burnups in figure 4–2 are expressed in megawatt-days (thermal) per kilogram of uranium (MWtd/kg U). In general, the design burnup of a reactor is inversely related to the initial enrichment level of the fuel. Data on burnup are not very useful for our purposes unless they are accompanied by data on enrichment levels.[29]

Table 4–1 presents empirical data on initial enrichment levels and actual burnup levels achieved in domestic reactors. Using these figures, the Nuclear Regulatory Commission has calculated duty factors, which we have used in turn to calculate fuel use, measured in short tons U_3O_8 in concentrate per net gigawatt-year (electrical). The figures for fuel use would have to be reduced 16.6 percent if a 0.20 percent tails assay were assumed. Unfortunately, the data source for table 4–1 does not report design burnup, so we cannot estimate the extent to which actual performance has fallen short of design objectives.

The federal government has been criticized for using duty factors based on design burnup in estimating uranium demand. Use of a duty factor of 14 million kWeh/ST U_3O_8 has been recommended by Day;[30] 18 million kWeh/ST U_3O_8 has been recommended by Kazmann and Selbin;[31] and 22 million kWeh/ST U_3O_8 has been recommended by Huntington, assuming 0.3 percent tails and 3.0 percent enrichment.[32] However, the Nuclear Regulatory Commission staff has defended the use of higher figures in testimony presented in reactor licensing hearings. Design burnup, rather than actual burnup, is used as a basis for calculations:

Fuel failure mechanisms that have resulted in the premature discharge of power reactor fuel assemblies have been discussed. Practically all of these have been due to three failure mechanisms: (1) hydriding, (2) pellet/cladding interaction and (3) fuel densification with cladding collapse. Suitable solutions have been developed to eliminate or minimize these failures. Hydride and fuel densification problems have been completely resolved. PCI problems are currently under review to supplement our understanding of the problem and confirm the effectiveness of the present solutions. These solutions, revisions to operating procedures and the use of smaller diameter fuel rod designs, should minimize failures due

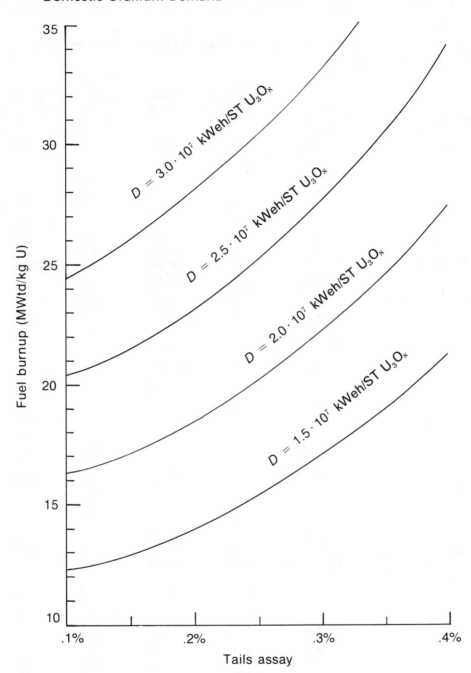

Figure 4–2. Duty Factors as a Function of Tails Assay and Burnup, Assuming an Enrichment Level of 3 Percent U-235 and No Recycle.

Table 4–1
Empirical Data on Duty Factors Achieved in Fuel Batches Removed from Domestic Reactors

Reactor	Batch Number	Initial Enrichment (Percent U-235)	Burnup[b] $\left(\dfrac{MWtd}{kg\ U}\right)$	Duty Factor[c] $\left(\dfrac{10^6\ kWeh}{ST\ U_3O_8}\right)$	Fuel Use[d] $\left(\dfrac{ST\ U_3O_8}{GWey}\right)$
Surry 1	1A1	1.87	14.971	22.9	383
	1A2	1.87	19.792	30.3	289
	2	2.57	22.584	23.9	367
Ginna	1	2.44	20.900	23.5	373
	2	2.78	18.689	18.1	484
	3	3.48	10.117	7.7	1138
	4A	3.22	25.135	20.7	423
	5A	2.26	19.318	23.7	370
PWR-1[a]	1	1.86	12.350	19.0	461
	2	2.56	22.650	24.1	363
H.B. Robinson	1	1.85	15.860	24.6	356
	2&3	2.825	25.100	23.7	370
	4	3.10	23.400	20.1	436
Point Beach 1	1	2.21	19.396	24.5	358
	2	3.04	30.755	27.0	324
	3A	3.40	13.177	10.2	859
	3B	3.40	25.344	19.6	447
Oconee 1	1	2.01	11.560	16.3	537
	2A	2.10	11.585	15.5	565
	2B	2.10	19.203	25.6	342
	3A	2.15	18.678	24.2	362
Maine Yankee	1	2.01	10.925	15.4	569
	2	2.40	10.911	12.5	701
	3	2.95	8.468	7.7	1138
	4	2.01	15.518	21.8	402
	5	2.40	15.993	18.3	479
	6	2.95	11.772	10.7	819
	7	1.93	5.111	7.5	·1168
	8	2.33	2.769	3.3	2655
Indian Point 1	1	2.80	12.655	12.2	718
	2	3.26	19.039	15.5	565
	3	4.08	24.215	15.4	569
	4	4.10	25.247	16.0	548
	5	4.08	22.884	14.6	600
Yankee Rowe	1	3.40	8.470	6.6	1327
	2	3.40	10.150	7.9	1109
	3	4.1	12.900	8.1	1081
	4	4.1	16.500	10.4	842
	5	4.1	17.250	10.9	804
Yankee Rowe	6	4.83	22.515	12.0	730
	7	4.94	24.282	12.6	695
	8	4.78	23.188	12.4	706
	9	4.94	23.874	12.4	706
	10	4.94	26.037	13.5	649
	11	4.00	25.277	16.4	534
Conn. Yankee	1	3.02	18.670	16.5	531
	2	3.24	25.035	20.5	427
	3	3.67	30.989	22.1	396
	4	3.65	27.685	19.9	440
	5	4.00	32.860	21.4	409

Table 4–1 continued

Reactor	Batch Number	Initial Enrichment (Percent U-235)	Burnup[b] $\left(\dfrac{MWtd}{kg\ U}\right)$	Duty Factor[c] $\left(\dfrac{10^6\ kWeh}{ST\ U_3O_8}\right)$	Fuel Use[d] $\left(\dfrac{ST\ U_3O_8}{GWey}\right)$
PWR-2[a]	1	3.16	17.500	15.2	576
	2	3.40	23.500	19.4	452
	3	3.86	28.900	20.1	436
	4	4.00	28.400	19.1	459
Quad Cities 1	1	2.13	9.360	12.9	679
	2	2.13	16.470	22.7	386
Quad Cities 2	1	2.13	11.490	15.9	551
	2	2.13	11.750	16.1	544
Dresden 2	1	2.13	12.413	17.1	512
	2	2.13	16.741	22.9	383
	3	2.30	3.253	4.1	2137
Dresden 3	1	2.13	6.899	9.5	922
	2	2.13	10.995	15.1	580
	3	2.13	13.936	19.2	456
Monticello	1	2.25	7.850	10.2	859
	2	2.25	13.750	17.9	489
	3	2.25	15.930	20.8	421
	4	2.25	16.900	21.9	400
Nine Mile Point 1	1	2.11	5.700	8.0	1095
	2	2.11	8.010	11.2	782
	3	2.11	12.580	17.5	501
	4	2.11	16.800	23.4	374
	5	2.11	18.950	26.4	332
	6	2.30	17.130	21.6	406
Oyster Creek	1	2.1	9.200	13.0	674
	2	2.1	12.200	17.1	512
	3	2.1	16.600	23.3	376
	4	2.1	19.800	27.8	315
	5	2.1	22.800	32.0	274
Oyster Creek	6	2.1	19.400	27.1	323
	7	2.6	10.400	11.4	768
Vermont Yankee	1	2.50	3.670	4.2	2086
	2	2.50	4.512	5.2	1685
	3	2.50	4.512	5.2	1685
	4	2.50	7.622	8.8	995
	5	2.50	9.124	10.4	842
	5A	2.50	.987	1.1	7964
	5B	2.50	9.124	10.4	842

Source: Philip M. Wood, "NRC Staff Testimony on Uranium Fuel Use Efficiency Presented by P.M. Wood," in the matter of Gulf States Utilities Company (River Bend Station, Units 1 and 2), May 1976.

[a]The name of the reactor has been withheld by the Nuclear Regulatory Commission at the request of the utility.

[b]Actual batch burnup.

[c]Assuming no recycle and 0.3 percent enrichment tails.

[d]Derived from duty factor; not reported in source of data.

to PCI. Thus, future fuel performance should be unaffected by these fuel failure mechanisms, and design fuel burnups for equilibrium cycles, typically 27,500 MWD/MT [megawatt-days (thermal) per metric ton] for a BWR and 33,000 MWD/MT for a PWR, should be readily achievable in practice.[33]

These burnup levels are associated with duty factors of 23 million kWeh/ST U_3O_8 in first cores and 34 to 36 million kWeh/ST U_3O_8 in subsequent cores for a boiling water reactor (BWR), and 30 to 34 million kWeh/ST U_3O_8 in first cores and 34 million kWeh/ST U_3O_8 in subsequent cores for a pressurized water reactor (PWR), assuming 0.3 percent tails and no recycle.[34]

In 1977 the Nuclear Regulatory Commission published a report which suggests that lower duty factors should be assumed for the purpose of making projections of future fuel requirements. Lifetime average annual fuel use for a "model" light-water reactor with a capacity of 1 GWe and a capacity factor of 80 percent was estimated to be 293 metric tons milled U_3O_8 per year, assuming 0.3 percent tails and no recycle.[35] This figure is equivalent to 21.7 million kWeh/ST U_3O_8; at 0.2 percent tails it corresponds to 26.0 million kWeh/ST U_3O_8.

In the remainder of this chapter and in chapter 6, we will need to make some assumption regarding the lifetime average duty factors which will be achieved by light-water reactors. However, we do not find that any particular assumption is strongly supported by empirical data. Certain fuel problems which have caused low duty factors in the past—hydriding and densification—appear to be resolved. However, other problems may crop up in the future, and duty factors may be reduced by reactor shutdowns and deratings associated with components other than fuel bundles. We prefer to take a somewhat pessimistic viewpoint and assume that 25 million kWeh/ST U_3O_8 will be achieved in practice. This corresponds to yellowcake requirements of 350 ST U_3O_8/GWey and uranium ore requirements of about 389 ST U_3O_8/GWey.

Price Competition with Coal and Oil

Once a duty factor has been estimated for a reactor, it is simple to derive a few figures related to price competition between uranium and coal and oil. The contribution of uranium concentrates to the busbar cost of nuclear power is measured in mills per kilowatt-hour:

$$\frac{P_u}{D} \cdot 2000 \text{ lb/ST} \cdot 1000 \text{ mills/\$} = \frac{2P_u}{10^{-6}D} \cdot \frac{\text{lb} \cdot \text{mills}}{\text{ST} \cdot \$} \qquad (4.6)$$

where P_u = price of uranium concentrate, in \$/lb U_3O_8

For example, if uranium costs \$37.50/lb U_3O_8 and a duty factor of 25 million kWeh/ST U_3O_8 is achieved, the yellowcake costs 3 mills/kWeh. The total busbar cost of nuclear power from a currently proposed plant may be on the order of 30 mills/kWeh, in 1977 dollars. For example, a capital cost of \$750/kWe with a 15 percent annual carrying charge and 60 percent capacity factor would result in a levelized capital cost of 21.4 mills/kWeh. To this we may add 3 mills/kWeh for U_3O_8, 4 mills/kWeh for enrichment and other fuel costs, and 2 mills/kWeh for operation and maintenance. These figures are intended only to show that the busbar cost of nuclear power is roughly ten times the cost of U_3O_8. Precise forecasts are impossible.[36]

It is instructive to compare the kinds of price changes in uranium and coal which will create an equal change in busbar cost. Let us assume that a change in nuclear power cost equals a change in coal-fired power cost:

$$\Delta P_u \cdot \frac{1}{D} \cdot 2000 \text{ lb/ST} \cdot 1000 \text{ mills/\$} \ = \ \Delta P_c \cdot \frac{H}{h_c} \cdot 1000 \text{ mills/\$} \quad (4.7)$$

where P_c = price of coal, in \$/ST coal
H = heat rate of coal-fired power plant, in Btu/kWeh
h_c = heat content of coal, in Btu/ST coal

If we assume a duty factor of 25 million kWeh/ST U_3O_8, a heat rate of 9,300 Btu/kWeh, and a heat content of 24 million Btu/ST coal, then every dollar per ton of increase in coal prices is equivalent to a \$4.84 per pound increase in U_3O_8 prices:

$$\frac{\Delta P_u}{\Delta P_c} = \frac{HD}{h_c \cdot 2000 \text{ lb/ST}}$$

$$= \frac{9.3 \cdot 10^3 \text{ Btu/kWeh} \cdot 2.5 \cdot 10^7 \text{ kWeh/ST } U_3O_8}{2.4 \cdot 10^7 \text{ Btu/ST coal} \cdot 2000 \text{ lb/ST}} \quad (4.8)$$

$$= \frac{\$4.84/\text{lb } U_3O_8}{\$/\text{ST coal}}$$

This figure would apply to a new coal-fired power plant burning Eastern coal. Similarly, if we consider a plant burning residual oil with a heat content of 5.8 million Btu/bbl, every dollar per barrel increase in oil prices is equivalent to a \$20.04 per pound increase in U_3O_8 prices:

$$\frac{\Delta P_u}{\Delta P_o} = \frac{HD}{h_o \cdot 2000 \text{ lb/ST}}$$

$$= \frac{9.3 \cdot 10^3 \text{ Btu/kWeh} \cdot 2.5 \cdot 10^7 \text{ kWeh/ST } U_3O_8}{5.8 \cdot 10^6 \text{ Btu/bbl} \cdot 2000 \text{ lb/ST}} \qquad (4.9)$$

$$= \frac{\$20.04/\text{lb } U_3O_8}{\$/\text{bbl}}$$

where P_o = price of residual oil, in \$/bbl
$\quad h_o$ = heat content of residual oil, in Btu/bbl

Finally, every cent per million Btu of increase in fossil fuel prices is equivalent to a \$1.16 per pound increase in U_3O_8 prices:

$$\frac{\Delta P_u}{\Delta P_f} = \frac{HD}{2000 \text{ lb/ST} \cdot 100 \text{ ¢/\$}}$$

$$= \frac{9.3 \cdot 10^3 \text{ Btu/kWeh} \cdot 2.5 \cdot 10^7 \text{ kWeh/ST } U_3O_8}{2000 \text{ lb/ST} \cdot 100 \text{ ¢/\$}} \qquad (4.10)$$

$$= \frac{\$1.16/\text{lb } U_3O_8}{\text{¢}/10^6 \text{ Btu}}$$

where P_f = price of fossil fuel, in ¢/10^6 Btu

In October 1976 the price of U_3O_8 for 1982 delivery rose to \$62 per pound, although old contracts had specified prices as low as \$8 per pound.[37] The Nuclear Exchange Corporation has reported that prices of \$11 per pound were quoted as recently as January 1973.[38] An increase of \$51 per pound from 1973 to 1976 may appear exorbitant, until one compares it with coal price increases. In July 1973 the New York State Electric and Gas Corporation estimated that its cost of coal in 1982 would be 70 cents per million Btu, but by March 1976 the estimate rose to 130 cents per million Btu.[39] An increase of 60 cents per million Btu would be equivalent to a \$69.75 per pound increase in U_3O_8 prices, assuming the burnup and heat rate previously given. Other coal-burning utilities face even greater changes in price expectation. These data suggest that uranium prices may rise far above the levels prevailing prior to 1973 without significantly reducing the quantity demanded. However, no definite statement on such price competition can be made without a thorough examination of nuclear and coal-fired power costs.

Reactor Lifetime Requirements

Some utilities have become concerned about ensuring the adequacy of fuel supply for a reactor over its useful lifetime. There are many ways in which this concern might be expressed:

1. A contract for delivery of lifetime requirements may be negotiated.
2. Reserves sufficient to cover these requirements may be purchased.
3. Land which is estimated to contain potential resources sufficient to cover these requirements may be purchased or the mineral rights acquired.
4. Long-term projections of uranium supply and demand may be developed.

In any case, it is necessary to calculate lifetime ore requirements per unit of nuclear capacity:

$$R_L = \frac{LC}{rD} \cdot 10^3 \text{ kWe/MWe} \cdot 365 \text{ d/y} \cdot 24 \text{ h/d}$$

$$= \frac{LC}{rD} \cdot 8.76 \cdot 10^6 \text{ kWeh/MWey} \tag{4.11}$$

$$= \frac{LC \, (e - t)}{rBE \, (1 - Z) \, (0.711 - t)} \cdot 473.48 \frac{\text{ST } U_3O_8 \cdot \text{kWed}}{\text{kg U} \cdot \text{MWey}}$$

where R_L = requirements for uranium ore reserves over the lifetime of a reactor, in ST U_3O_8/MWe
   L = lifetime of a reactor, in y
   C = capacity factor, averaged over the reactor's lifetime (a fraction)
   r = recovery rate; the fraction of U_3O_8 contained in ore reserves, which is recovered in the form of yellowcake
   MWe = megawatt (electrical)
   y   = year

Given L, C, and r, we can express R_L as a function of tails assay and burnup. A commonly used value for L is 30 years, although the useful life of a reactor is determined by economic factors which are difficult to forecast over such a long time frame. A capacity factor of 60 percent would represent a slight improvement over recent operating experience. Recovery rates have been declining over the past 10 years; for the next 30 years, we may assume an average value of 90 percent. Recovery rates will be discussed in more detail in chapter 5; my estimate is based upon figure 5–8 and equation 5.12. Using these assumptions for L, C, and r, lifetime requirements have been plotted in figure 4–3. At a duty factor of 25 million kWeh/ST U_3O_8, lifetime ore requirements are 7.008 ST U_3O_8 per net MWe of capacity.

It is important to note that capacity factor and burnup are related. If nuclear fuel is defective and must be replaced, values for both these variables will be decreased. One difficulty with estimating a relationship between the two, in order to project future values, is that experts such as the Nuclear Regulatory Commission staff claim that past experience with fuel burnup provides no indication of future performance. Although the discrepancy between actual and currently projected capacity factors is debated more widely, it is relatively minor when compared to the discrepancy between actual and currently projected burnup rates.

The Nuclear Regulatory Commission's estimate of uranium concentrate requirements for the "average" LWR (a weighted average of BWR and PWR), at a 0.30 percent tails assay, is 6.357 ST U_3O_8/MWe.[40] Given a 90 percent recovery rate, this is associated with ore requirements of 7.063 ST U_3O_8/MWe. This estimate is basically in agreement with my preceding estimate, because the Nuclear Regulatory Commission assumes higher capacity factors as well as higher burnup levels. If a tails assay of 0.20 percent were assumed instead of 0.30 percent, ore requirements would be 5.302 ST U_3O_8/MWe.

Slightly higher values, at a 0.20 percent tails assay, are suggested by data published by the Atomic Energy Commission in 1971. A rough estimate of lifetime ore requirements can be obtained simply:

$$R_L = \frac{I + (L_n - 1)M}{r} \qquad (4.12)$$

where I = uranium concentrate required to fuel an initial core, in ST U_3O_8/MWe

M = uranium concentrate required to supply the net annual makeup, in ST U_3O_8/MWe

L_n = number of years in a reactor lifetime

The relevant data, based on a 0.20 percent tails assay, are shown in table 4–2. A reduction in duty factor, below design specifications, would be consistent with these data as long as a proportionate reduction in

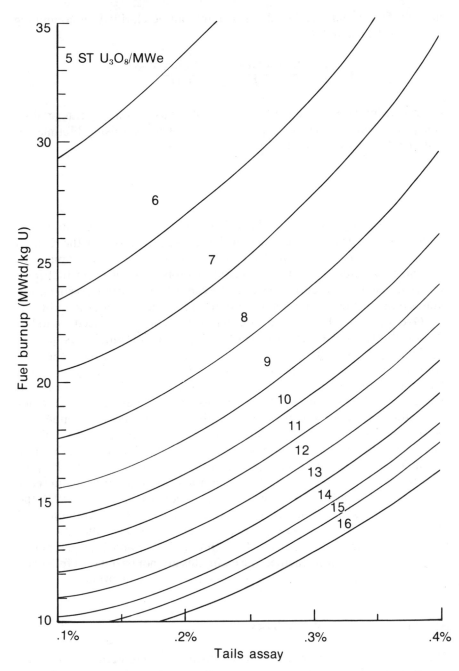

Figure 4–3. Reactor Lifetime Requirements as a Function of Tails Assay and Burnup, Assuming an Enrichment Level of 3 Percent U-235 and No Recycle.

capacity factor were achieved. This point can be clarified by combining equations 4.11 and 4.12:

$$\frac{D}{C} = \frac{L \cdot 8.76 \cdot 10^6 \text{ kWeh/MWey}}{I + (L_n - 1)M} \qquad (4.13)$$

The replacement of defective fuel can easily cause a much greater reduction in duty factor than in capacity factor, thereby raising lifetime requirements above the values shown in table 4–2.

Fuel Recycling

All the preceding calculations have been based on the assumption that nuclear fuel will not be recycled in light-water reactors in the United States. In other words, we have assumed that recycled uranium or plutonium will not be used in domestic reactors, even if spent fuel from domestic reactors is reprocessed, and even if foreign reprocessing companies offer recycled uranium or plutonium for sale to domestic utilities.

Over the past 10 years, the federal government has released a number of projections for the fuel recycling industry, each of which was quickly replaced by a subsequent projection. In December 1967, plutonium recycling was scheduled to be introduced in 1974.[41] By February 1974, the date had been advanced to 1977,[42] and by April 1977, it had been advanced to 1990.[43] At this point ERDA adopted President Carter's recommendation that domestic fuel reprocessing be deferred indefinitely because of hazards associated with plutonium.[44] The economic and political outlook for domestic plutonium recycle is now sufficiently bleak that I do not anticipate any significant impact on uranium requirements to the year 2000. Without plutonium recycle, the economic outlook for uranium recycle is also very bleak. However, we may note that in August 1976 the Nuclear Regulatory Commission claimed that "if both uranium and plutonium are recycled, the U_3O_8 requirements could be reduced by 22.3 percent over the 26-year period (1975 through 2000) and 29.3 percent in the year 2000."[45] These engineering estimates are not based on a full-scale demonstration program, so they are not necessarily accurate.

Scheduling of Uranium Purchases

The preceding discussion has laid the basis for a projection of uranium requirements for a typical light-water reactor. In order to describe the quantity of uranium demanded in the marketplace, one must examine

Table 4–2
U_3O_8 Requirements for Particular Reactors (Design Specifications)

Reactor	Feed Required per MWe, in Tons U_3O_8, Assuming 0.2 Percent Tails		
	Initial Core (I)	Net Annual Makeup (M)	$R_L = \dfrac{I + 29M}{0.9}$
BWRs:			
Dresden 2	0.842	0.150	5.769
Pilgrim	0.867	0.152	5.861
Quad Cities 1	0.850	0.157	6.003
Browns Ferry 1	0.708	0.163	6.039
Peach Bottom 2	0.727	0.158	5.899
Fitzpatrick	0.687	0.147	5.500
Ringhals 1	0.786	0.166	6.222
			Mean 5.899
PWRs:			
Turkey Point 3	0.603	0.178	6.406
Salem 1	0.548	0.172	6.151
Mihama 1	0.798	0.174	6.493
Oconee 1	0.591	0.156	5.683
Prairie Island 1	0.615	0.183	6.580
Kewaunee	0.636	0.177	6.410
Zion 1	0.540	0.171	6.110
Cook 1	0.535	0.166	5.943
Maine Yankee	0.606	0.181	6.506
Calvert Cliffs 1	0.612	0.176	6.351
Arkansas One	0.616	0.154	5.647
Trojan	0.466	0.170	5.996
Farley 1	0.519	0.186	6.570
Biblis	0.523	0.167	5.962
San Onofre 2	0.453	0.161	5.691
			Mean 6.167
Model BWR, starting up in 1976–1980	0.680	0.145	5.428
Model BWR, starting up in 1981–1985	0.635	0.140	5.217
Model PWR, starting up in 1976–1980	0.545	0.165	5.922
Model PWR, starting up in 1981–1985	0.480	0.165	5.850

Source: Values of *I* and *M* can be found in USAEC, *Forecast of Growth of Nuclear Power*, WASH-1139 (Washington: USAEC, January 1971), pp. 17–18.

more than these technical data—one must consider the scheduling of uranium purchases. The topic of purchasing behavior is complex and imprecise; it involves not only the study of actual price fluctuations, but also the study of long-term price expectations. I will defer the discussion of prices to chapters 7 and 8, and therefore the present discussion will be brief.

Production Lead Times

The primary constraint on uranium purchasing is the relationship between reactor and fuel-cycle lead times. At present, these are both about 10 years, as table 4–3 shows. However, the present market is influenced by government policies established in the 1960s, when reactor lead times were expected to be much shorter. Since utilities were not expected to make firm uranium purchase commitments until they had ordered reactors, the gap in lead times was seen as a justification for government intervention in the marketplace. The Atomic Energy Commission purchased large quantities of uranium during the 1960s and accumulated a stockpile of 50,000 short tons in excess of its own requirements, on the assumption that a free market would fail to sustain uranium production at a rate needed to meet nuclear capacity growth projections.

The estimated reactor lead time of 10 years is really only a lower bound in the sense that it is contingent upon electricity demand projec-

Table 4–3
Estimates of Lead Times

From	To	1964 Estimate[a] (years)	1976 Estimate (years)
Decision to explore	Establishment of reserves	4–6	5[b]
Establishment of reserves	Production of U_3O_8	4–6	3[b]
Production of U_3O_8	Initial reactor operation	1	1.75[c]
Total fuel cycle		9–13	9.75
Decision to build	Initial reactor operation	4	10[d]

[a]Allan E. Jones, "Remarks Presented at the 9th Annual Minerals Symposium, Uranium Section, Moab, Utah, May 22, 1974"; reprinted in U.S. Congress, Joint Committee on Atomic Energy, *Private Ownership of Special Nuclear Materials, 1964,* Hearings, 88th Cong., 2d Sess., June 9, 10, 11, 15, and 25, 1964 (Washington: GPO, 1964), p. 179.

[b]Lead times for establishment of reserves are idealized; in actuality, a range exists. See John Klemenic, "Analysis and Trends in Uranium Supply," *Uranium Industry Seminar,* GJO-108(76) (Grand Junction, Colo.: USERDA, 1976), p. 232.

[c]The estimate of 1.75 years between U_3O_8 production and initial reactor operation assumes no stockpiling. See Ralph M. Rotty, A.M. Perry, and David B. Reister, *Net Energy from Nuclear Power,* IEA-75-33 (Oak Ridge, Tenn.: Institute for Energy Analysis, 1975).

[d]The reactor lead time assumes current trends in licensing and construction. See U.S. General Accounting Office, *Reducing Nuclear Powerplant Leadtimes: Many Obstacles Remain,* EMD-77-15 (Washington: USGAO, March 2, 1977), pp. 2, 15.

tions. Delays in reactor construction schedules may be caused by reductions in demand projections. An increase in demand projections, however, cannot cause a shortening of lead times.

Advance Coverage

Some utilities contract for uranium many years in advance, while others prefer only short-term contracts. All utilities are obligated to establish an "assured" supply of fuel for any new power plants they construct, but this term can be interpreted so loosely that it is of no use in studying utility contracting behavior. One utility executive has described the situation in the following terms:

> Assurance is determined by the individual utility, who in purchasing a nuclear power plant must decide if there is a high enough probability that he [*sic*] will be able to acquire fuel supply for the plant over its 30 year life. "Assurance" is a rather vague term since it means different things to different people. In one case, it may mean a twenty year supply of uranium in inventory while in another case, it could be based upon the fond hopes of an exploration program.[46]

Empirical data on uranium supply arrangements are shown in table 4–4. It should be clear from these data that the idea of a "twenty year supply of uranium in inventory" is totally farfetched. If we assume that fuel reloads are scheduled annually, then it can be seen from table 4–4 that by May 1977, uranium supplies sufficient to operate a reactor for 10 or more years had been arranged for only 23 percent of the nuclear capacity for which nuclear steam supply systems had been ordered. In other words, 77 percent of the capacity in operation, under construction, or on order did not have enough uranium under contract to keep the reactor running for as long as 10 years.

Short-term coverage reached a high in 1971; medium-term coverage reached a high in 1977; and overall coverage reached a low in 1976. The 1976 values are strongly influenced by the announcement by Westinghouse Electric Corporation on September 8, 1975 that it would not honor its uranium supply contracts, due to "commercial impracticability." This situation will be discussed in chapter 7.

Spot Purchases

In every year except 1970, deliveries of uranium concentrate from domestic producers to domestic buyers have fallen short of the official

Table 4-4
Percentages of Nuclear Generating Capacity Covered by Uranium Supply Arrangements, 1968–1977

Reload	June 1968	May 1969	May 1970	May 1971	April 1972	April 1973	April 1974	April 1975	April 1976	May 1977
1st core	80	86	84	86	76	61	68	61	61	71
1	70	73	70	73	66	55	62	58	53	66
2	62	66	68	67	60	48	54	50	43	57
3	53	59	60	61	51	42	50	44	34	49
4	38	39	42	44	39	29	38	37	27	44
5	27	26	29	34	28	21	32	31	22	39
6	19	21	18	23	17	15	24	23	11	33
7	7	8	8	11	10	11	19	18	9	28
8	5	6	5	10	8	9	14	14	6	24
9		5	4	7	7	4	10	13	5	23
10			3	6	5	3	7	11	5	22
11			1	2	3	2	4	11	3	18
12				2	3	2	4	9	3	15
13				2	3	2	4			
14				1	2	1	3			
15					1	1	3			
16							3			
17							3			
18							2			
19							2			
20							1			

Source: Annual AEC and ERDA surveys of U.S. uranium marketing activity.

Note: For first cores, the data pertain to reactors for which a nuclear steam supply system has been ordered, but no operating license has been granted. For reloads, reactors with operating licenses are also included. A reload corresponds to roughly 1 year of reactor operation. A blank space indicates that data are unavailable.

projections made in the previous year. This trend is shown in table 4–5. Although one conclusion to be drawn from these data is that official government projections should be regarded with skepticism, it is also reasonable to conclude that the spot market represents only a small proportion of uranium deliveries. However, it is possible that spot market prices will have a major impact on average uranium prices through the mechanism of market price contracting—a development which will be discussed in chapter 7.

Inventories

Data on uranium inventories are presented in table 4–6. Producers' inventories have been principally "pipeline" inventories, equivalent to only

Table 4–5
Comparison of Scheduled Deliveries of Yellowcake (to Domestic Buyers)
with Actual Deliveries
(*in thousands of short tons* U_3O_8)

	Domestic Supplies		Imports	
Year	Deliveries Scheduled at Beginning[a] of Year	Actual Deliveries	Deliveries Scheduled at Beginning[a] of Year	Actual Deliveries
1966–1967		0.9		
1968	5.3[b]	4.8		
1969	4.6	4.2		
1970	8.9[c]	9.3		
1971	12.8[c]	12.7		
1972	11.7	11.6		
1973	12.3	12.1		
1974	13.7	11.9		
1975	15.6	12.5	0.8	1.1
1976	15.9	13.8	2.8	2.9
1977	15.9	13.9	4.0	2.8[d]
1978	19.1		1.6[d]	

Source: Annual AEC, ERDA, and DOE surveys of U.S. uranium marketing activity, 1968–1978.

[a]January 1, unless otherwise noted.

[b]Scheduled as of April 1.

[c]Scheduled as of February 1.

[d]These figures do not include uranium that has been re-exported or is committed to be re-exported, or material that is under litigation.

2 to 5 months of deliveries. Buyers' inventories have grown to almost 2 years' worth of deliveries. These data tend to refute the thesis, advanced by Vince Taylor and others, that the federal government has created a short-term scarcity of uranium by increasing uranium concentrate stockpiles above their economically optimal level via enrichment contracting policies.[47]

Nuclear Capacity Projections

Next, we may turn to the question of what kind of growth in domestic nuclear capacity can be anticipated. Past trends are shown in figure 4–4. Projections of installed nuclear capacity in 1980 are shown in table 4–7 and figure 4–5; projections for the year 2000 are shown in table 4–8 and figure 4–6. It is clear that a sharp decline in projections and in capacity on order (but not under construction) has occurred in recent years.

Table 4–6
Inventories of Uranium
(in thousands of tons U$_3$O$_8$)

Date	Held by Domestic Producers			Domestic Concentrate and UF$_6$ Held by Domestic Buyers	Imported Concentrate and UF$_6$ Held by Domestic Buyers
	In ore at Mills	*In Process at Mills*	*Concentrate at Mills*		
1/1/71	0.145	0.419	3.401	7.4[a]	
1/1/72	0.196	0.467	2.100	8.6	
1/1/73	0.271	0.468	3.701	14.4	
1/1/74	0.113	0.328	5.238	19.9	
1/1/75	0.284	0.440	4.302	20.2	
1/1/76			3.3	22.6	1.1
1/1/77			2.3	25.8	3.5
1/1/78			2.5[b]	25.1	3.6

Source: Walter C. Woodmansee, "Uranium" in *Minerals Yearbook* for 1971 through 1974, issued by U.S. Department of the Interior, Bureau of Mines (Washington: GPO); and U.S. Energy Research and Development Administration, *Survey of United States Uranium Marketing Activity* for 1976 and 1977 (Washington: USERDA), and U.S. Department of Energy, *Survey of United States Uranium Marketing Activity* for 1978 (Washington: USDOE).

[a]Excludes UF$_6$.
[b]Includes UF$_6$.

Table 4–7
Estimates of Domestic Nuclear Capacity Installed in 1980
(in gigawatts)

Date	Report	Low	Median	High
Dec. 1962	"Civilian Nuclear Power"		40	
March 1965	WASH-1055	61		92
June 1966	Press Release S-20-66	80	95	110
Dec. 1967	WASH-1084	120	145.5	170
March 1968	WASH-1082		150	
May 1969	(Unpublished)		149	
Jan. 1971	WASH-1139		150	
Dec. 1971	WASH-1139 (Rev.1)	132	151	165
Dec. 1972	WASH-1139 (72)	127	131.6	144
Feb. 1974	WASH-1139 (74)	85.0	102.1	112.4
Feb. 1975	WASH-1139 (74) update	70	76 82	110
Oct. 1975	"Uranium Requirements"	70	76	82
March 1976	"Demand for Uranium"	70		76
Oct. 1976	"Uranium Requirements"	60	67	71
April 1977	"Changing Nuclear Picture"		61	68

Note: The complete references for these reports can be found in the section of the bibliography pertaining to projections of installed nuclear capacity in the United States.

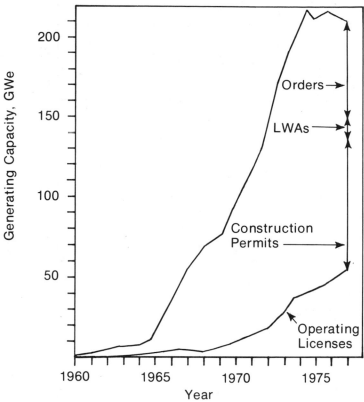

Source of data: Atomic Industrial Forum, Inc., "Profile of U.S. Nuclear Power Development," 1978.

Figure 4–4. United States Nuclear Power Capacity.

Table 4–8

Estimates of Domestic Nuclear Capacity Installed in 2000

(*in gigawatts*)

Date	Report	Low	Median	High
Dec. 1976	WASH-1084		735	
March 1968	WASH-1082		900	
May 1971	"Outlook for Uranium"	1200		1500
Dec. 1972	WASH-1139 (72)	825	1200	1500
Feb. 1974	WASH-1139 (74)	850	1090 1200	1400
Feb. 1975	WASH-1139 (74) update	625	800 1000	1250
Oct. 1975	"Uranium Requirements"	625	800	1000
March 1976	"Demand for Uranium"	625		800
Oct. 1976	"Uranium Requirements"	380	510	620
April 1977	"Changing Nuclear Picture"		380	500

Note: The complete references for these reports can be found in the section of the bibliography pertaining to projections of installed nuclear capacity in the United States.

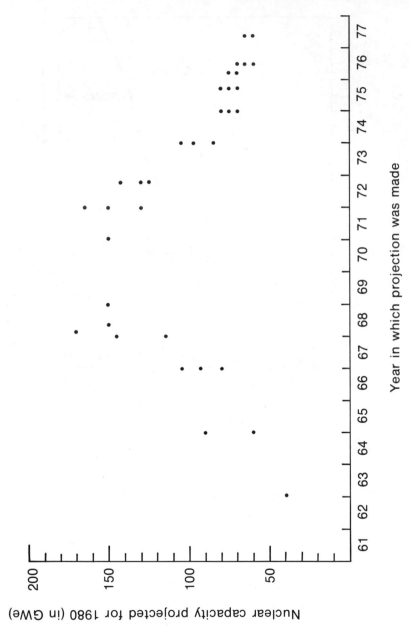

Figure 4–5. Nuclear Capacity Projected for 1980.

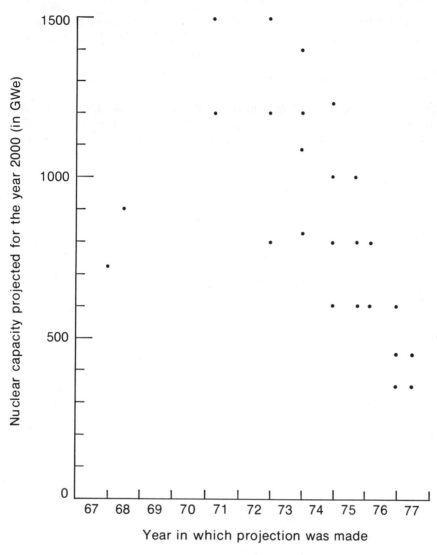

Figure 4–6. Nuclear Capacity Projected for the Year 2000.

At present, the pattern of growth in nuclear capacity covered by operating licenses, construction permits, and limited work authorizations provides evidence that installed nuclear capacity may reach a plateau of about 175 GWe in 1995. We consider this to be a sensible assumption for a low projection—a ceiling of 175 GWe, with no growth beyond that level. If present trends do not continue, however, nuclear capacity could rise above this level in the late 1990s, if not sooner. We believe that it is very unlikely for installed capacity to exceed 300 GWe by the year 2000.

Uranium Demand Projections

Finally, we may look at official forecasts of domestic uranium requirements. Forecasts for the entire United States are shown in figure 4–7; forecasts for the 208 GWe under ERDA enrichment contracts are shown in

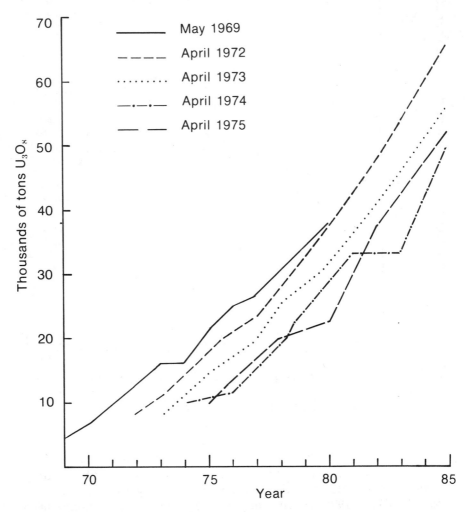

Figure 4–7. Official Forecasts of Domestic Uranium Requirements, Published in the Annual AEC/ERDA Surveys of Uranium Marketing Activity.

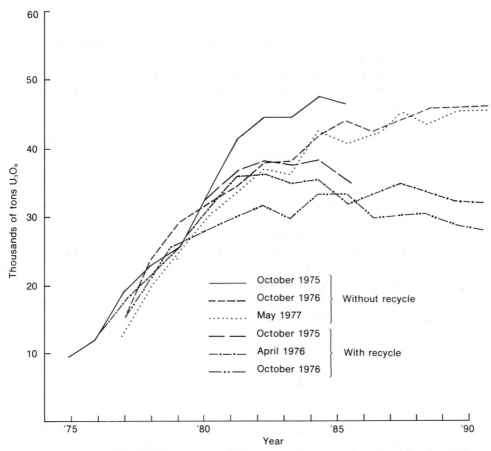

Figure 4–8. Official Forecasts of Uranium Required to Supply the 208 GWe under ERDA Enrichment Contracts.

figure 4–8. The overall pattern is a decline associated with declining reactor capacity projections followed by an upward shift due to the abandonment of fuel reprocessing. The variability of official forecasts indicates that the most recent projections should not be accepted as totally reliable; allowances should be made for a band of uncertainty.

Since 1974 there have been numerous delays and cancellations of nuclear power plants. The fuel supply contracts associated with these reactors have not necessarily been delayed and canceled proportionately. Among the contracts which are relatively inflexible are "fixed commitment" contracts for DOE enrichment services. It should be noted that the "excess" of enrichment services is not very large when viewed as a time lag between nuclear capacity projections and enriched uranium projections.

Notes

1. U.S. Atomic Energy Commission, *Nuclear Power Growth 1974– 2000*, WASH-1139 (74) (Washington: USAEC, 1974), p. 25.

2. Ralph M. Rotty, A.M. Perry, and David B. Reister, *Net Energy from Nuclear Power*, IEA-75-33 (Oak Ridge, Tenn.: Institute for Energy Analysis, 1975).

3. H. Mandel, "Uranium Demand and Security of Supply—A Consumer's Point of View," in *Uranium Supply and Demand*, edited by M.J. Spriggs (London: Uranium Institute, 1976), p. 83.

4. "Cancellations Pick Up As Growth Rates Remain Stable," *Not Man Apart* 5,19 (October 75):16.

5. "Florida Power Corp. Gives Up on Nuclear Power for the 1980's," *Nucleonics Week*, December 25, 1975.

6. Statement by William Walbridge, General Manager of the Sacramento Municipal Utility District, quoted in "SMUD Tables Rancho Seco 2," *Not Man Apart* 6,3 (February 76):9.

7. "Calendar of Procedural Steps for Operational Approval of U.S. Power and/or Experimental Reactors," *Nuclear Safety* 16,6 (November-December 75):764–770.

8. "Virginia Electric Says 2 Nuclear Units Canceled by Board," *Wall Street Journal*, March 21, 1977, p. 12.

9. "Utility Suits Pile Up," *Nuclear Industry*, October 1975.

10. Paul L. Joskow and Martin L. Baughman, "The Future of the U.S. Nuclear Energy Industry," *Bell Journal of Economics* 7,1 (Spring 1976):18–19, 29.

11. Edward J. Hanrahan, Richard H. Williamson, and Robert W. Brown, "The Changing Nuclear Picture: Uranium and Separative Work Requirements," paper presented at the Atomic Industrial Forum Fuel Cycle Conference 1977, Kansas City, Mo., April 25, 1977.

12. Ibid., p. 14.

13. *ERDA News* 2,18 (September 5, 1977):6.

14. Joskow and Baughman, "The Future of the U.S. Nuclear Energy Industry," p. 19.

15. R. Gene Clark and Andrew W. Reynolds, "Uranium Market— Domestic and Foreign Requirements," *Uranium Industry Seminar*, GJO-108(78) (Grand Junction, Colo.: USDOE, in press, 1978), Figure 10.

16. Atomic Industrial Forum, *Nuclear INFO*, September 1978, p. 4.

17. See note 12 in chapter 2.

18. This term was coined by Vince Taylor. See Vince Taylor, *How the U.S. Government Created the Uranium Crisis (and the Coming Uranium Bust)* (Los Angeles, Calif.: Pan Heuristics, June 1977), p. 17.

19. "U.S. Urged to Set Up Independent Firm for Nuclear Fuel," *Wall Street Journal*, July 21, 1977.

20. Joe W. Parks and David C. Thomas, "Plans for Operating Enrichment Plants and the Effect on Uranium Supply," *Uranium Industry Seminar*, GJO-108(76) (Grand Junction, Colo.: USERDA, 1976), p. 25.

21. 31 F.R. 16479 (December 23, 1966); and 32 F.R. 16289 (November 29, 1967).

22. Martin Becker, "Prepared Testimony before the New York State Board on Electric Generation and the Environment," in the matter of Long Island Lighting Company—Jamesport Nuclear Units 1 and 2 (1977), p. 89. For a graph of feed/product ratios and SWU/product ratios as a function of tails assay, see B.H. Cherry, "Uranium: The Market and Adequacy of Supply," in *Mineral Resources and the Environment, Supplementary Report: Reserves and Resources of Uranium in the United States*, prepared by the Committee on Mineral Resources and the Environment of the National Research Council (Washington: National Academy of Sciences, 1975), p. 223.

23. U.S. Energy Research and Development Administration, *Statistical Data on the Uranium Industry*, GJO-100(77) (Grand Junction, Colo: USERDA, 1977), p. 14; Walter C. Woodmansee, "Uranium," in *Minerals Yearbook, 1970*, issued by U.S. Department of the Interior, Bureau of Mines (Washington: GPO, 1972), p. 1146.

24. Victor K. McElheny, "Major Nuclear Role for Gas Centrifuge," *New York Times*, April 20, 1977.

25. Parks and Thomas, "Plans for Operating Enrichment Plants," pp. 15–16.

26. "Centrifuge, Cascade Advances to Double Enrichment Capacity," *ERDA News* 2,15 (July 25, 1977):1.

27. "EEI Surveys Utilities, Proposes Guidelines," *Nuclear News*, July 1977, pp. 43–45.

28. Philip M. Wood, "NRC Staff Testimony on Uranium Fuel Use Efficiency Presented by P.M. Wood," in the matter of Gulf States Utilities Company (River Bend Station, Units 1 and 2), p. 15.

29. We do not know of a publication which regularly issues these data.

30. M.C. Day, "Nuclear Energy: A Second Round of Questions," *Bulletin of the Atomic Scientists* (December 1975):54.

31. Raphael G. Kazmann and Joel Selbin, Letter to the Editor, *Scientific American* 234,4 (April 1976): 8–10. See also Raphael G. Kazmann, "Do We Have a Nuclear Option?" *Mining Engineering* (August 1975): 35–37.

32. John J. Berger, *Nuclear Power: The Unviable Option* (Palo Alto, Calif.: Ramparts, 1976), p. 160.

33. Dean Houston, "Testimony Regarding Light Water Reactor Fuel Performance Prepared by D. Houston," in the matter of Pacific Gas and Electric Company (Diablo Canyon Nuclear Power Plant Unit Nos. 1 and 2), p. 7.

34. Philip M. Wood, "NRC Staff Testimony," pp. 17–18.

35. U.S. Nuclear Regulatory Commission, *Environmental Survey of the Reprocessing and Waste Management Portions of the LWR Fuel Cycle*, NUREG-0116 (Washington: USNRC, 1977), p. 3-14.

36. The Atomic Industrial Forum surveys current nuclear generating costs annually. See "Nuclear Power Ends Best Performance Year Ever; Production in '77 Up 31%, Generation Costs Stable," INFO news release, April 20, 1978.

37. U.S. Energy Research and Development Administration, *Survey of United States Uranium Marketing Activity*, ERDA 77-46 (Washington: USERDA, 1977), p. 8; "Pennsylvania P & L Buys UNC Uranium at $56 in 1980 and $62 in 1982," *Nucleonics Week*, October 28, 1976.

38. U.S. Congress, Joint Committee on Atomic Energy, *Proposed Modification of Restrictions on Enrichment of Foreign Uranium for Domestic Use*, Hearings 93d Cong., 2d Sess., September 17–18, 1974 (Washington: GPO, 1975), pp. 40–41.

39. *1973 Report of the New York State Electric and Gas Corporation Pursuant to Article VIII, Section 149-B of the Public Service Law*, Vol. 1, Exhibit 9, p. 9; *1977 Report of Member Electric Systems of the New York Power Pool and the Empire State Electric Energy Research Corporation Pursuant to Article VIII, Section 149-B of the Public Service Law*, Vol. 1, p. 393.

40. P.M. Wood, "NRC Staff Testimony," Table 24.

41. U.S. Atomic Energy Commission, *Forecast of Growth of Nuclear Power*, WASH-1084 (Washington: USAEC, December 1967), p. 10.

42. U.S. Atomic Energy Commission, *Nuclear Power Growth, 1974–2000* WASH-1139(74) (Washington: USAEC, February 1974), p. 33.

43. Hanrahan, Williamson, and Bown, "The Changing Nuclear Picture," p. 11.

44. U.S. Energy Research and Development Administration, *LWR Spent Fuel Disposition Capabilities*, ERDA 77-25 (Washington: USERDA, May 1977), p. 1.

45. U.S. Nuclear Regulatory Commission, *Final Generic Environmental Statement on the Use of Recycle Plutonium in Mixed Oxide Fuel in Light Water Cooled Reactors*, NUREG-002 (Washington: USNRC, August 1976), Vol. 2, p. III-42.

46. B.H. Cherry, "Uranium: The Market and Adequacy of Supply," in National Academy of Sciences, Committee on Mineral Resources

and the Environment, *Mineral Resources and the Environment—Supplementary Report: Reserves and Resources of Uranium in the United States* (Washington: National Academy of Sciences, 1975), p. 224.

47. Taylor, *How the U.S. Government Created the Uranium Crisis*, p. 78.

5 Uranium Production Costs

There are many difficulties involved in estimating uranium production costs. Three areas of concern may be listed. First, it is difficult to generalize about different types of deposits and different production methods. Two methods of open-pit mining and nine methods of underground mining may be distinguished.[1] We will simply distinguish open-pit from underground mining. Second, it is difficult, if not impossible, to obtain up-to-date empirical data on costs experienced by domestic uranium producers. Nearly all published data involve estimates and projections for hypothetical cases. I have not attempted to check these estimates against company records, which are proprietary. And third, many important parameters of uranium production are affected by uranium prices, and future prices are likely to remain higher than prices experienced during the 1963–1973 period. It is difficult to predict exactly how the technology will adapt to higher prices.

Given these concerns, we would like to preface the following discussion by noting that this investigation is largely exploratory, and some inaccuracies may be revealed by future research on this topic. Our principal objectives in developing a cost model are threefold:

1. To provide information needed to estimate revenues from federal leasing programs, given a variety of assumptions about severance taxes, royalties, and federal income taxes
2. To describe the way in which the grade of ore milled will vary in response to uranium prices and costs
3. To determine an optimal production capacity for a given mining situation.

Discussion of production costs is divided into four parts. First, published data will be assimilated in order to estimate some basic cost equations. Second, the basic analytical model will be presented, to show how these equations can be used. Next, a brief discussion of "forward costs"—the basis for official reserve estimates—is presented. Finally, information is supplied on three highly uncertain cost components: exploration costs, land reclamation, and stabilization of uranium mill tailings.

111

Interpretation of Published Data

Capital and Operating Costs

To our knowledge, the most detailed published analyses of uranium production costs are those issued by the Grand Junction Office of the Department of Energy. Since estimates of domestic uranium reserves and
resources are prepared annually on a "forward cost" basis rather than on
the basis of purely geological parameters, cost estimation is an integral
part of the resource evaluation program. We may distinguish several
components of forward cost for a hypothetical uranium deposit:[2]

1. Capital costs
   a. Property acquisition
   b. Field expense, exploration drilling, and development drilling
   c. Mine primary development, mine plant and equipment, and mill
      construction
2. Operating costs
   a. Royalty and severance (ad valorem) tax
   b. Mining, hauling, and milling costs

Our interest, initially, is in 1c and 2b. For a known uranium deposit, one
must deduct from this list any capital costs which have already been
incurred; forward costs of reserves must therefore be defined in terms of a
specific date, such as January 1 of a given year.

In the accounting system used by the Department of Energy, capital
costs, like operating costs, are expressed in terms of dollars per pound of
U_3O_8 in concentrate. However, no return on investment and no federal
income taxes are included. Rather, these capital costs represent the ratio
of capital expenditure (in constant dollars) to the number of pounds of
U_3O_8 in concentrate ultimately recovered from a deposit. For our purposes, these figures must be converted to expenditures per unit of production capacity.

Historically, the average grade of uranium ore milled in the United
States has been above 0.15 percent U_3O_8, that is, at least 3 pounds of U_3O_8
per short ton of ore.[3] The most recent study of production costs for such
ores was released in 1972,[4] so we must escalate costs to 1977 levels. Table
5–1 shows the indexes used, and tables 5–2 and 5–3 show the results.

Capital Expenditures. Assume that the annual production rate of ore is
constant over the production lifetime of a hypothetical mining and milling
complex. We have found no discussion in published literature of the
accuracy of this assumption, and therefore there is no evidence that a

Table 5-1
Indexes of Cost Escalation for Uranium Production

Cost Element	Type of Index^a	Industry Group	June 1972 Value	June 1977 Value	Change (%)
Capital costs					
Mine primary development and mine plant and equipment:					
Open-pit mines	WPI	Scrapers and graders	124.9	220.6	76.6
Underground mines	WPI	Mining machinery and equipment	117.6	226.5	92.6
Mill construction	WPI	Machinery and equipment	118.1	196.0	66.0
Operating costs					
Mining: Labor	AHE	Metal mining	4.40	7.29	65.7
Fuel	WPI	Petroleum products, refined	108.5	311.6	187.2
Composite					126.4
Hauling: Labor	AHE	Trucking & trucking terminals	4.85	7.09	46.2
Fuel	WPI	Petroleum products, refined	108.5	311.6	187.2
Composite					116.7
Milling: Labor	AHE	Industrial inorganic chemicals	4.46	6.83	53.1
Materials	WPI	Basic inorganic chemicals	108.7	189.8	74.6
Composite					63.9

Source: U.S. Department of Labor, Bureau of Labor Statistics, *Wholesale Prices and Price Indexes*, June 1977 (Washington: GPO, August 1977) and June 1972 (Washington: GPO, August 1972); *Employment and Earnings*, August 1977 (Washington: GPO, August 1977) and August 1972 (Washington: GPO, August 1972).

^aWPI = wholesale price index, relative to 1967 values (which equal 100); AHE = average hourly earnings, in $/h; percentage changes in the composite indexes are estimated simply by averaging the changes in AHE and the WPI.

Table 5-2
Cost Estimates for Uranium Production from Open-Pit Mines with a Depth-to-Thickness Ratio of 24 and an Ore Grade of 0.20 Percent U$_3$O$_8$
(in $/ST ore, in 1977 dollars)

	Capacity (ST ore/d)				
	500	*1,000*	*2,000*	*3,000*	*5,000*
Capital costs					
Mine primary development	10.77	9.54	9.18	9.09	8.92
Mine plant and equipment	0.35	0.35	0.35	0.35	0.35
Mill construction	3.98	3.24	2.66	2.32	1.99
Total	15.10	13.13	12.19	11.76	11.26
Operating costs					
Mining	5.43	5.43	5.43	5.43	5.43
Hauling	1.41	1.41	1.41	1.41	1.41
Milling	9.92	7.62	6.31	5.65	5.41
Total	16.76	14.46	13.15	12.49	12.25

Source: Indexes in table 5-1 have been applied to data from John Klemenic, "Examples of Overall Economics in a Future Cycle of Uranium Concentrate Production for Assumed Open Pit and Underground Mining Operations" (Grand Junction, Colo.: USAEC, 1972), p. 2.

Table 5-3
Cost Estimates for Uranium Production from Underground Mines with a Depth-to-Thickness Ratio of 76 and an Ore Grade of 0.25 Percent U$_3$O$_8$
(in $/ST ore, in 1977 dollars)

	Capacity (ST ore/d)				
	500	*1,000*	*2,000*	*3,000*	*5,000*
Capital costs					
Mine primary development	7.99	6.26	5.30	5.01	4.53
Mine plant and equipment	1.73	1.35	1.06	0.96	0.87
Mill construction	3.98	3.24	2.66	2.32	1.99
Total	13.70	10.85	9.02	8.29	7.39
Operating costs					
Mining	31.70	27.17	24.90	23.77	22.87
Hauling	1.73	1.73	1.73	1.73	1.73
Milling	10.16	7.87	6.56	5.90	5.65
Total	43.59	36.77	33.19	31.40	30.25

Source: Indexes in table 5-1 have been applied to data from John Klemenic, "Examples of Overall Economics in a Future Cycle of Uranium Concentrate Production for Assumed Open Pit and Underground Mining Operations" (Grand Junction, Colo.: USAEC, 1972), p. 7.

more complicated production profile would be more realistic. It is simple to compute the total quantity of ore mined:[5]

$$R = q \cdot F \cdot T \cdot 365 \text{ d/y} \qquad (5.1)$$

where R = quantity of ore mined over the production lifetime, in ST ore
$\quad q$ = installed daily capacity of the mine-mill complex, in ST ore/d
$\quad F$ = capacity factor at which ore production is maintained
$\quad T$ = production lifetime, in y
$\quad d$ = day
$\quad y$ = year

Capital expenditures per unit of production capacity can be found as follows:

$$C = \frac{I \cdot R}{q} = I \cdot F \cdot T \cdot 365 \text{ d/y} \qquad (5.2)$$

where C = capital expenditures per unit of installed capacity, in $/ST ore/d
$\quad I$ = capital costs, expressed as the ratio of capital expenditures to the quantity of ore mined, in $/ST ore

Data on the aggregate capacity factor of domestic uranium mills are shown in table 5–4. Hogerton et al. assume 300 day per year operation for new facilities, corresponding to an 82 percent capacity factor.[6] We choose to assume an 80 percent capacity factor, the average of the 1976 performance and the projection by Hogerton et al. The data in tables 5–2 and

Table 5–4
Aggregate Capacity Factor of Domestic Uranium Mills

Year	Ore Fed to Process (10^6 ST ore)	Nominal Mill Capacity (10^3 ST ore/d)	Capacity Factor
1974	6.987	26.65	0.72
1975	7.5	28.45	0.72
1976	8.9	31.16	0.78
1977	10.1	39.21	0.71

Source: U.S. Energy Research and Development Administration, *Statistical Data of the Uranium Industry,* GJO-100(75) (Grand Junction, Colo.: USERDA, 1975), pp. 78, 80; GJO-100(76), pp. 80, 82; and GJO-100(77), pp. 100, 102; U.S. Department of Energy, Statistical Data of the Uranium Industry, GJO-100(78) (Grand Junction, Colo.: USDOE, 1978), pp. 80–81.

5–3 are based on a production lifetime of 10 years.[7] Given these assumptions, we have fitted cost functions to the data in table 5–2, using regression analysis, and estimated capital costs for open-pit production:[8]

$$C_D = 2.583 \cdot 10^4 + 7.729 \cdot 10^{10} \, (\ln q)^{-9} \qquad (5.3)$$

$$C_E = 1022 \qquad (5.4)$$

$$C_M = -9.878 \cdot 10^3 + 1.336 \cdot 10^5 \, (\ln q)^{-1} \qquad (5.5)$$

where C_D = capital expenditures for mine primary development, in \$/ST ore/d
C_E = capital expenditures for mine plant and equipment, in \$/ST ore/d
C_M = capital expenditures for mill construction, in \$/ST ore/d

Similarly, the data in table 5–3 can be used to describe underground production:

$$C_D = 9.350 \cdot 10^3 + 2.073 \cdot 10^7 \, (\ln q)^{-4} \qquad (5.6)$$

$$C_E = 8.675 \cdot 10^2 + 1.004 \cdot 10^6 \, (\ln q)^{-3} \qquad (5.7)$$

Capital costs of mill construction are the same in both cases, since open-pit ores are indistinguishable from underground ores.

The schedule of these expenditures is shown in figure 5–1. Mine equipment is assumed to have a 5-year life.[9] Open-pit mine development ceases 2 years before the end of the production lifetime.

The data in tables 5–2 and 5–3 are based on ore grades of 0.20 percent and 0.25 percent, respectively. As of January 1, 1977, the average grade of domestic reserves at a forward cost no greater than \$30 per pound U_3O_8 was 0.09 percent.[10] According to a study published by the Grand Junction Office of the Atomic Energy Commission in 1974, capital costs per short ton of ore tend to decline as ore grade declines, in the range of 0.10 to 0.01 percent.[11] However, costs for mine primary development, mine plant and equipment, and mill construction combined were estimated to be higher for 0.10 percent ores in 1974 dollars than for 0.20 or 0.25 percent ores in 1972 dollars. We shall assume that in the range of 0.10 percent and above, the effect of ore grade on capital expenditures per unit of capacity is negligible. Cost estimates for ores significantly below 0.10 percent are rather speculative, and we shall avoid modeling such costs. Note that all these ore grades are average ore grades. The model does incorporate low values for marginal (minimum) ore grades.

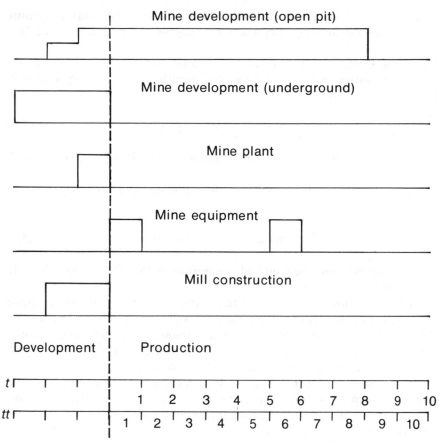

Source: John Klemenic, "Examples of Overall Economics in a Future Cycle of Uranium Concentrate Production for Assumed Open Pit and Underground Mining Operations" (Grand Junction, Colo.: USAEC, 1972), pp. 2, 7; John Klemenic, "An Estimate of the Economics of Uranium Concentrate Production from Low Grade Sources" (Grand Junction, Colo.: USAEC, 1974), pp. 3, 7, 20.

Figure 5-1. Schedule of Capital Expenditures.

In calculating capital expenditures per unit of capacity, we have implicitly assumed that this variable tends to be more stable over a range of capacities and deposit sizes than the ratio of capital expenditures to the quantity of ore mined. We do not know whether this assumption is valid when applied to mine primary development. In the case of open-pit mines, we have developed a computing procedure to calculate the amount of annual expenditure for mine primary development during the production period, assuming a 10-year mine life, and to apply this level of expenditure to all mines of equal capacity regardless of mine life. This procedure allows open-pit development costs to vary as a function of mine life.

Operating Costs. We have fitted a cost function to the data in table 5–2, using regression analysis, to estimate operating costs for open-pit production:

$$K = 10.99 + 5.366 \cdot 10^4 \, (\ln q)^{-5} \qquad (5.8)$$

where K = operating cost of mining, hauling, and milling ore, in $/ST ore

Similarly, we have used table 5–3 to estimate costs for underground production:

$$K = 26.82 + 1.559 \cdot 10^5 \, (\ln q)^{-5} \qquad (5.9)$$

These equations must be modified to incorporate the effect of variations in ore grade.

Mill operating costs per short ton of ore tend to increase as ore grade increases. In other words, chemical processing costs depend not only on the amount of ore feed, but also on the amount of yellowcake product. The data we have used show a 15 cent per ton cost differential (in 1972 dollars) between 0.20 and 0.25 percent ores, an 18 cent per ton differential (in 1974 dollars) between 0.05 and 0.10 percent ores, and a 16 cent per ton differential (in 1974 dollars) between 0.01 and 0.05 percent ores.[12] We escalate these figures to a 25 cent per ton differential (in 1977 dollars) per 0.05 percent change in ore grade.

Hauling costs per short ton of ore are totally unaffected by ore grade. Hauling costs are principally affected by the distance between the mine and the mill, but we choose not to model different assumptions about this distance. In 1968, when uranium prices were lower than they are today, it was reported that "since the cost of transporting U_3O_8 as ore is high, a mill usually is located within 15 to 20 miles of major ore sources, although this distance varies directly with the grade of ore and the cost of mining."[13] Commonwealth Edison owns an unusually high-grade uranium deposit, from which ores are transported 125 miles to be milled.[14]

Mine operating costs per ton of ore tend to increase as ore grade increases. Obviously, the cost of extracting the particular ton of ore which gets hauled to the mill has nothing to do with ore grade. However, high-grade ores involve the mining of a larger amount of waste rock and low-grade material (which does not get hauled to the mill) per ton of ore hauled than low-grade ores. High-grade ores tend to occur in thin deposits and small pockets, surrounded by less valuable material.

Part of the waste and low-grade material is mined simply because it is difficult to locate high-grade material, not because the ore is scattered in locations which are hard to reach. Mining practices at the Jackpile-Paguate mines, New Mexico, one of the world's largest open-pit uranium mines, have been described as follows:

> Eventually the material is mined from blocked out areas and segregated into waste, low grade, and ore with the ore going to stockpiles conforming to established categories. Shipments to the mill are controlled by blending from the stockpiles to obtain the desired ore grade for any given period.[15]

Subsurface drilling can be conducted to locate ores. Improvements in drilling technology, and other methods of locating uranium, will tend to reduce the need for mining and scanning waste.

In evaluating mine operating costs, we face two contradictory sets of data. Forward cost data, in dollars per pound of U_3O_8 recovered, are available for ores in the 0.15 to 0.30 percent range.[16] After exploration costs and royalties (based on 0.20 percent ore for open-pit production and 0.25 percent ore from underground production) are deducted, and the costs are converted to dollars per short ton of ore (in 1972 dollars),[17] we find roughly a 40 cent per ton differential between 0.15 and 0.20 percent ore, a 20 cent per ton differential between 0.20 and 0.25 percent ore, and a 20 cent per ton differential between 0.25 and 0.30 percent ore. Given the difference in mill operating cost discussed earlier, these data leave little room for variations in mine operating cost. However, wide variations are reported for ores in the 0.01 to 0.10 percent range, with much lower costs for 0.10 percent ores (in 1974 dollars) than for 0.20 or 0.25 percent ores (in 1972 dollars).[18] Using the data given for a 5,000 ton per day capacity, we looked for a function which would achieve a compromise between the two sets of data by indicating that costs for 0.10 percent ore were underestimated and costs for 0.15 percent ore were overestimated. We found that a very simple function, the square root of ore grade, would do the job, as figure 5–2 shows. We have no reason to reject the functional form on theoretical grounds, so we will use it. These curves are based upon 1972 dollars; an upward shift is necessary to estimate costs in 1977 dollars.

The equation for operating costs for open-pit mines is as follows:

$$K = 10.99 + 5.366 \cdot 10^4 \, (\ln q)^{-5} + 5 \, (G_a - 0.20) + 13 \, (\sqrt{G_a} - 0.45)$$

$$(5.10)$$

where G_a = average grade of ore, in percent U_3O_8.

For underground production, the equation is similar:

$$K = 26.82 + 1.559 \cdot 10^5 \, (\ln q)^{-5} + 5 \, (G_a - 0.25)$$
$$+ 45 \, (\sqrt{G_a} - 0.5) \qquad (5.11)$$

In each case, the final term models the effect of ore grade upon mine operating costs. We have estimated a relationship to the average grade of ore, although theoretically it would be better to estimate a relationship to the cutoff grade of ore, that is, the minimum grade of ore hauled to the mill. Data on cost implications of cutoff grades are unavailable, to my knowledge.

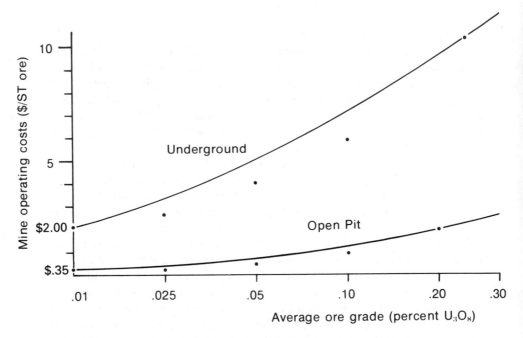

Note: The curves are the basis for the final term in equations 5.10 and 5.11. The curves refer to 1972 dollars, and the equations refer to 1977 dollars.

The data points at ore grades of 0.20 and 0.25 percent are estimates by Klemenic (1972) which we accept because they are based on empirical data. See figure 5–8 for historical levels of the average grade of uranium ore processed.

The data points at lower ore grades are estimates by Klemenic (1974) and are plotted for comparative purposes. We assume that costs were underestimated for 0.025 percent, 0.05 percent, and 0.10 percent ores. We accept the cost estimates for 0.01 percent ores but assume that they can only be rough approximations.

Figure 5–2. Estimation of the Effect of Average Ore Grade on Mine Operating Costs, at a Capacity Level of 5,000 Short Tons of Ore Per Day.

Recovery Rates

Uranium reserves and resources are usually measured in terms of the amount of ore or the amount of U_3O_8 in ore which can be mined. In other words, a 100 percent rate of recovery of ore is assumed by definition. In a given area, the total amount of rock containing significant amounts of uranium may be referred to as a "deposit," or "mineralization," but not "ore." For example, "The deposit could contain 20 million tons. An ore volume [*sic*] of 10 million tons will be realized at 50 percent mine efficiency."[19]

The recovery rate of U_3O_8 from ore is never 100 percent; it declines as ore grade declines. Ideally, the recovery rate should be set at the level where the marginal cost of yellowcake equals the marginal revenue. However, published data do not provide us with a basis for estimating the effect of prices on recovery rates. In a study prepared by the Grand Junction Office in 1974, recovery rates of 0.925 from 0.10 percent ore, 0.875 from 0.05 percent ore, 0.775 from 0.025 percent ore, and 0.475 from 0.01 percent ore were assumed.[20] These data can be described by a continuous function:

$$x = 0.9321 + 2.217 \cdot 10^{-4} (\ln G_a)^5 \qquad (5.12)$$

For 0.15 percent ore, a recovery rate of 0.927 is predicted by this equation. This is consistent with 1976 industry experience.[21] For earlier years, the equation underestimates industrywide recovery rates slightly, but we do not consider this a significant problem in the context of modeling future production. For ore grades below 0.01 percent U_3O_8, this equation predicts very low recovery rates, but we are not too concerned about underestimating such rates because cutoff grades are likely to exceed 0.01 percent in sandstone deposits. Even if it becomes profitable to mine sandstone deposits at ore grades below 0.01 percent, such operations might be prohibited in the future by environmental restrictions.

For a given grade of ore, it is technically possible to increase the recovery rate by crushing the ore to a finer mesh standard, by increasing the amount of acid or other leaching solution used per ton of ore, and by recycling waste streams. Obviously, the design of milling machinery and equipment can be modified in response to price changes. We have not found data which would permit us to estimate such a price response.

Mineral Inventories

For the simulation model, it is desirable to include among the data inputs some information about the geological characteristics of a uranium deposit, independent of costs and prices. In particular, we need to know the way ore grade declines as the quantity of ore mined increases. Data on preproduction mineral inventories for the United States, New Mexico, and Wyoming were published for the first time in 1977.[22] Figure 5–3

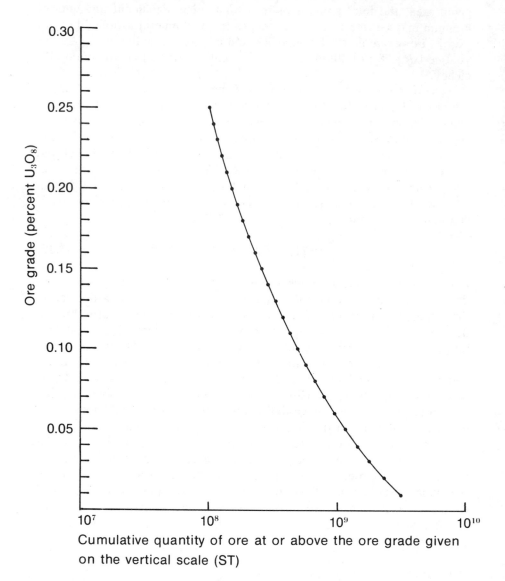

Figure 5–3. United States Preproduction Uranium Mineral Inventory, January 1, 1977.

illustrates the relationship between ore grade and quantity for the United States. The curve can be described by the following equation:

$$G = 3.059\, R_M^{-0.5447} - 0.03 \qquad (5.13)$$

where G = grade of ore, in percent U_3O_8
  R_M = quantity of ore present at grades equal to or higher than G, in 10^6 ST ore

New Mexico's uranium inventory can be described similarly:

$$G = 2.084\, R_M^{-0.5162} - 0.05 \qquad (5.14)$$

Data for Wyoming also fit a curve closely:

$$G = 1.134\, R_M^{-0.4831} - 0.03 \qquad (5.15)$$

Accordingly, we cite as a general case,

$$G = a \cdot R^b - c \qquad (5.16)$$

where R = quantity of ore present at grades equal to or higher than G, in ST ore

Obviously, the parameters a, b, and c may vary widely from one deposit to another. We do not have empirical data on the ore grade distributions for individual deposits.[23] Without such information, it is impossible to model the way average grade and cutoff grade vary in response to U_3O_8 prices and production costs. Consequently, we will use hypothetical values of a, b, and c for the stimulation model, and evaluate different patterns of uranium deposits by changing these parameters.

Development of a Basic Analytical Model

Our next task is to outline a basic model, that is, to describe the way the preceding cost data can be used. We believe the presentation will be clarified by first examining the case of fixed prices and costs, and then relaxing this assumption.

Fixed Prices and Costs

Let us consider first the situation in which production capacity q is an input to the model. An iterative procedure must be used to determine the average grade of ore G_a. Given values for q and G_a, other variables, including the production lifetime, can be determined.

The iterative procedure must begin with some assumption about G_a. A value of $G_a^* = 0.10$ percent may be selected. Operating costs can then be calculated using equation 5.10 or equation 5.11, and the recovery rate can be calculated using equation 5.12. The cutoff grade of ore is determined by equating the marginal cost per short ton of ore to the marginal revenue per short ton of ore:

$$K = (1 - \Lambda)(1 - s)P\frac{G_c}{100\%}x \cdot 2000 \text{ lb/ST}$$

$$- [\delta + \phi \, (1 - \delta)]\left[(1 - z)(1 - \Lambda)(1 - s)P \, \frac{G_c}{100\%}x \cdot 2000 \text{ lb/ST} - K\right]$$

$$(5.17)$$

where δ = state income tax rate
Λ = royalty rate, in relation to P
s = severance tax rate, in relation to P
P = price of uranium concentrate, in \$/lb U_3O_8
G_c = cutoff grade of ore, in percent U_3O_8
ϕ = federal income tax rate
z = depletion rate, in relation to revenue

This equation can be solved for G_c:

$$G_c = \frac{\{1 - [\delta + \phi \, (1 - \delta)]\} \, K \cdot 100\%}{\{1 - [\delta + \phi \, (1 - \delta)](1 - z)\}(1 - \Lambda)(1 - s)Px \cdot 2000 \text{ lb/ST}} \qquad (5.18)$$

Then, equation 5.16 can be solved for R_c, the quantity of ore to be mined:

$$R_c = \left(\frac{G_c + c}{a}\right)^{\frac{l}{b}} \qquad (5.19)$$

where R_c = quantity of ore presented at grades equal to or higher than the cutoff grade, in ST ore

The average grade of ore can be found by integration using equation 5.16. Because of the nature of the curve, the lower limit of integration must not be zero, so a small number may be chosen (see figure 5–4):

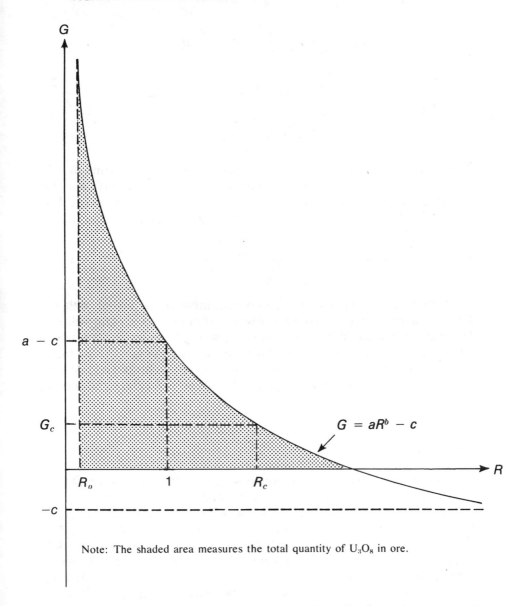

Figure 5–4. General Relationship Between Ore Grade and Amount of Ore Reserve.

$$G_a = \frac{1}{R_c - R_o} \int_{R_0}^{R_c} (ar^b - c)\, dr$$

$$= \frac{1}{R_c - R_o} \left(\frac{a}{b+1} r^{b+1} - cr \right) \Big|_{R_0}^{R_c} \qquad (5.20)$$

$$= \frac{a\,(R_c^{b+1} - R_o^{b+1})}{(R_c - R_o)\,(b+1)} - c$$

Then, a new value of G_a^* may be selected—the average of G_a and the old value of G_a^*—and the calculation of operating cost and so forth may be repeated. A solution is found when G_a is sufficiently close to G_a^*. The production lifetime can then be calculated, given equation 5.1:

$$T = \frac{R_c}{Fq \cdot 365 \text{ d/y}} \qquad (5.21)$$

Finally, let us suppose that q is not an input to the model; rather it must be selected in order to maximize the after-tax net present value of the uranium mining lease. Since the derivation of T, given q, is rather complicated, it is best to maximize q by trying out a range of values of q rather than starting with a range of values of T. By contrast, oil and coal production can be modeled by trying out a range of T values and solving for q.[24]

Variable Prices and Costs

Let us begin again with the situation in which q is taken as an input to the model. The production lifetime can be divided into discrete time intervals of 1 year each. The average grade can be determined for the first interval, then the second, and so forth until reserves are exhausted.

Let tt be an index of discrete time (see figure 5–1). For a given interval, an iterative procedure must be followed similar to the one previously described. A value of $G_a^* = 0.10$ may be selected initially, and the operating cost and recovery rate may be calculated. The cutoff grade involves price and cost escalation:

$$G_c(tt) = \frac{\{1 - [\delta + \phi(1 - \delta)]\}Ke^{\theta i} \cdot 100\%}{\{1 - [\delta + \phi(1 - \delta)](1 - z)\}(1 - \Lambda)(1 - s)Pe^{\Omega t}x \cdot 2000 \text{ lb/ST}}$$

$$(5.22)$$

where θ = index of operating cost escalation
    Ω = index of price escalation

The next step is to calculate the quantity of ore which would be mined if the cutoff grade $G_c(tt)$ were maintained throughout the production lifetime:

$$R_c(tt) = \left(\frac{G_c(tt) + c}{a} \right)^{\frac{1}{b}} \qquad (5.23)$$

The average grade can be calculated as in equation 5.20 and a new value of G_a^* established for the next iteration, as before. Once $G_a(tt)$ has been estimated, the amount of ore mined in year tt can be expressed as a fraction of the total amount which could be mined at $G_c(tt)$:

$$e(tt) = \frac{Fq \cdot 365 \text{ d/y}}{R_c(tt)} \qquad (5.24)$$

There are two ways in which the final year of the production lifetime TT can be reached:

If $\quad \displaystyle\sum_{s=1}^{tt} e(s) = 1 \quad$ then $TT = tt$

If $\quad \displaystyle\sum_{s=1}^{tt} e(s) > 1 \qquad\qquad (5.25)$

then $TT = tt \quad$ and $\quad e(TT) = 1 - \displaystyle\sum_{s=1}^{TT-1} e(s)$

In the latter case, equation 5.24 must be solved for the value of F in the final time interval. An important aspect of this modeling procedure is that it incorporates what is known as "high-grading"—when a high cutoff grade is used in a given year, ores below that grade tend to become "lost" to further mining.

Finally, let us suppose that q values must be selected to maximize after-tax net present value of the lease. A range of q values must be tried and the results compared. We will return to a discussion of the analytical model in chapter 9.

Forward Cost of U.S. Reserves

One of the deficiencies in the production cost information previously described is that it does not permit a range of assumptions regarding the depth and thickness of uranium deposits. The Grand Junction Office has published data on mine primary development costs as a function of the depth/thickness ratio, based on empirical data from the 1960s.[25] We have chosen not to use these data because they are limited to mine primary development, because depth and thickness are not treated individually, and because the data set is rather old. However, some insight into the effects of depth and thickness on production cost can be obtained from information on current U.S. reserves.

The amount of U_3O_8 estimated to be minable from known reserves at or below certain marginal forward costs is illustrated in figure 5–5. The data for reserves at $10, $15, and $30 per pound U_3O_8 fit the following relationship:

$$M_u = 5.271 \ e^{2.556Q} \qquad (5.26)$$

where M_u = marginal forward cost of U.S. reserves, in $/lb U_3O_8
  Q = quantity of U.S. reserves minable at costs equal to M_u or lower, in 10^6 ST U_3O_8

The average forward cost of $30 reserves can be determined by integration:

$$A_u = \frac{1}{0.68} \int_0^{0.68} 5.271 \ e^{2.556Q} \ dQ = \$14.21/\text{lb} \ U_3O_8 \qquad (5.27)$$

where A_u = average forward cost of U.S. reserves, in $/lb U_3O_8

Average forward costs per short ton of ore can be found quite easily:

$$A_o = A_u \cdot \frac{G}{100\%} \cdot x \cdot 2{,}000 \ \text{lb/ST} \qquad (5.28)$$

where A_o = average forward costs of U.S. reserves, in $/ST ore

Fourteen classes of deposits are described in table 5–5. We have calculated A_o for each class, assuming a value of $14.21/lb U_3O_8 for A_u and a recovery rate of 0.92. In other words, we have assumed that the ratio of

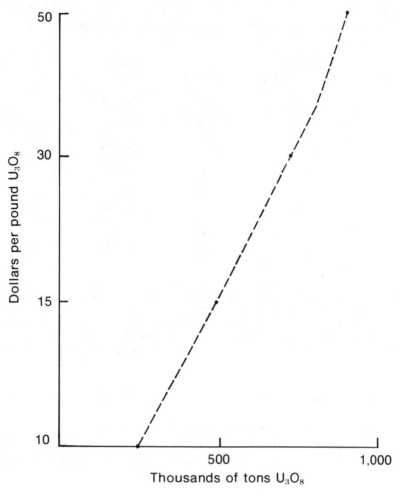

Figure 5–5. U.S. Uranium Reserves, January 1, 1977.

average forward cost to marginal forward cost is the same for each class of deposit as it is for the aggregate.

We have obtained an ordinary least squares fit to these data as follows:

$$A_o = 4.64 + 6.50\,D_1 + 16.30\,D_2 + 2.44\,T_1$$
$$(0.57)\quad(2.29)\qquad(5.28)\qquad(0.93)$$

$$+\ 5.89T_2 + 67.75(\ln S)^{-1}\quad r^2 = 0.67 \qquad (5.29)$$
$$(1.35)\qquad(1.24)$$

Table 5-5
Statistics on 14 Classes of Reserve Deposits, January 1, 1977[a]

Reserve (10^3 tons U_3O_8)	Average Grade (percent U_3O_8)	Average Cost ($/ton ore)	Average Size of Deposit (10^3 tons ore)	Depth Range (feet)	Thickness Range (feet)
337.07	0.10	26.15	5436.58	500+	8+
105.92	0.07	18.31	5043.76	250–500	8+
45.33	0.06	15.69	5811.69	250–500	0–8
42.57	0.06	15.69	3941.22	0–250	8+
24.04	0.15	39.23	1780.79	500+	0–8
24.03	0.07	18.31	2288.68	0–250	0–8
16.63	0.12	31.38	532.92	500+	8+
14.05	0.06	15.69	433.64	0–250	8+
13.87	0.09	23.54	395.18	250–500	8+
9.93	0.07	18.31	315.19	0–250	4–8
8.11	0.10	26.15	311.75	250–500	4–8
6.13	0.12	31.38	319.30	500+	0–8
4.17	0.07	18.13	250.36	0–250	0–4
0.64	0.13	34.00	247.53	250–500	0–4
652.49					

Source: U.S. Energy Research and Development Administration, *Statistical Data of the Uranium Industry*, GJO-100(77) (Grand Junction, Colo.: USERDA, 1977) p. 67.
[a]Each class is defined by a deposit size range (either 100–1,000 ST ore or more than 1,000 ST ore), a depth range, and a thickness range. All these reserves are available at forward costs no greater than $30/lb U_3O_8. Deposits minable by open-pit and underground methods are aggregated in each class.

where D_1 = dummy variable for depth of deposit between 250 and 500 feet

D_2 = dummy variable for depth of deposit over 500 feet

T_1 = dummy variable for thickness of deposit between 4 and 8 feet

T_2 = dummy variable for thickness of deposit between 0 and 4 feet

S = average size of deposit, in 10^3 ST ore

The *t* statistics are shown in parenthesis. The data exhibit a reasonably good fit to the regression model, considering the fact that these forward costs do not include capital expenditures made prior to January 1, 1977.

The implications of this regression model are illustrated in table 5-6. We have selected production capacities of 500 and 3,000 tons per day for comparison with tables 5-2 and 5-3 and estimated deposit sizes by assuming a 75 percent capacity factor and 7-year production lifetime and by using equation 5.1. In general these forward costs are well below the sum of capital and operating costs in the corresponding columns of tables 5-2 and 5-3. The data suggest that depth is a more important variable than thickness, and that consideration of both variables is preferable to the use of a depth/thickness ratio.

Table 5–6
Estimated Average Forward Cost of Uranium Production from U.S. Reserves
(*in 1977 dollars per short ton of ore*)

Capacity (ST ore/d)	Depth Range (feet)	Thickness Range (feet)		
		8+	4–8	0–4
500	0–250	14.51	16.95	20.40
500	250–500	16.37	18.81	22.26
500	500+	30.81	33.25	36.70
3,000	0–250	12.47	14.91	18.36
3,000	250–500	14.33	16.77	20.22
3,000	500+	28.77	31.21	34.66

Source: These costs have been calculated using equations 5.1 and 5.29.

Other Production Costs

Exploration Cost

In evaluating leasing alternatives, we will assume that the permit to develop the land is issued after uranium reserves have been identified. This assumption is consistent with leasing practice on public lands withdrawn for use of the Atomic Energy Commission, on acquired lands, and on certain Indian lands (for example, the Navajo-Exxon lease). We may ask, therefore, what exploration costs are likely to have been incurred by the time a deposit of a given size has been identified. Any estimate of these costs must be highly uncertain. Using historical data, we can develop a rough, short-term projection of the average exploration expenditure per pound of U_3O_8 added to reserves on a nationwide scale. However, we do not have data regarding the extent to which costs for an individual deposit may vary from this average.

Costs per pound U_3O_8 discovered can be found by calculating the ratio of cost per foot drilled to pounds U_3O_8 discovered per foot drilled. This may appear to be unnecessarily cumbersome, but it is common practice. If the number of pounds U_3O_8 discovered per foot drilled could be defined in purely geological terms, it would be desirable to separate projections of this "discovery rate" from projections of economic variables. However, the available data on discoveries are expressed in terms of the amount of U_3O_8 which is either added to reserves on a forward cost basis or mined and shipped to mills in the same calendar year in which it is discovered. We do not consider discovery rates based on these data to be an "objective" measure of the success of industry exploration programs.

The topic is addressed here because of the widespread attention it has received in published literature.

There are several ways in which costs per foot drilled, and discovery rates, can be defined. We choose to develop a projection of the following ratio, in the second form in which it is expressed:

$$\frac{E_e + E_d + E_o}{A} = \frac{\dfrac{E_e + E_d + E_o}{D_e + D_d}}{\dfrac{A}{D_e + D_d}} = \frac{\dfrac{E_e + E_d}{D_e + D_d}}{\dfrac{A}{D_e + D_d}} + \frac{E_o}{A}$$

$$= \frac{\dfrac{E_e}{D_e}}{\dfrac{A}{D_e}} + \frac{\dfrac{E_d}{D_d}}{\dfrac{A}{D_d}} + \frac{E_o}{A}$$

(5.30)

where E_e = expenditures for exploration drilling in a given year, in constant dollars

E_d = expenditures for development drilling in a given year, in constant dollars

E_o = other exploration expenditures, excluding land acquisition, incurred in a given year, in constant dollars

D_e = surface exploration drilling in a given year, in feet

D_d = surface development drilling in a given year, in feet

A = following year's additions to constant-dollar reserves plus cumulative production from mined ore, in pounds U_3O_8

There are four different expressions for costs per foot drilled in the numerators of equation 5.30; note that our projection must be based on constant dollars. Three of these measures are graphed in figure 5–6, in current dollars.

We have selected a definition of A which enables us to make use of data published by the Energy Research and Development Administration in 1976.[26] These data are based on constant-dollar reserves, which are estimated by interpolating (for each year) along a curve of the type illustrated in figure 5–5. The curve is based on current dollars, and interpolations are made after the current-dollar equivalent of some cost level (say, $15 per pound U_3O_8, in 1975 dollars) has been calculated according to the wholesale price index for industrial commodities.

The various components of an annual change in current-dollar reserves are listed in table 5–7. The figure used for discovery rates is the constant-dollar equivalent of the third from the bottom; it has been derived by working backwards from the bottom line, after the latter has been measured on a constant-dollar basis. The data are shown in the first

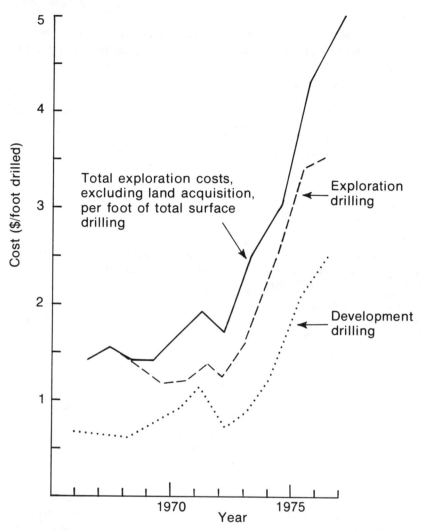

Source: These figures have been calculated from data in William L. Chenoweth, "Exploration Activities," *Uranium Industry Seminar*, GJO-108(76) (Grand Junction, Colo.:
USERDA, 1976), pp. 180, 183; and U.S. Energy Research and Development Administration, *Statistical Data of the Uranium Industry*, GJO-100(77) (Grand Junction, Colo.:
USERDA, 1977), pp. 75–79.

Figure 5–6. Costs per Foot of Drilling, 1966–1976.

column of table 5–8. Note that there are two difficulties with these data:
(1) production from ore is not on a constant-dollar basis at all; and (2) it
would be better to derive constant-dollar reserves on the basis of a
uranium cost index rather than the wholesale price index for industrial
commodities. Discovery rates are shown in table 5–8. We define *A* in

Table 5–7

Components of an Annual Change in Current-Dollar Uranium Ore Reserves, According to the Accounting System Used by the Department of Energy

	1977 Values for $30 Reserves ($10^3$ ST U_3O_8)
Additions found on new properties	39.0
Plus additions associated with reevaluation of old properties for which reserves had been underestimated	+62.0
Minus subtractions associated with the shifting of reserves into higher-cost categories, due to inflation, or the reevaluation of old properties for which reserves had been overestimated	−46.0
Minus subtractions due to erosion, that is, mining at a cutoff grade which omits some of the reserves	− 1.0
Change in current-dollar reserves plus cumulative production[a]	54.0
Minus production from mine water, solution mining, heap leaching, and unrecorded low-grade stockpiles	− 0.5
Additions to current-dollar reserves plus cumulative production (Klemenic-Sanders)[b]	53.5
Minus production from ore	−13.5
Change in current-dollar uranium ore reserves	40.0

Source: U.S. Energy Research and Development Administration, *Statistical Data of the Uranium Industry*, GJO-100(77) (Grand Junction, Colo.: USERDA, 1977), pp. 12, 24, 28.

[a]The "sum of reserves and cumulative production" is published annually in *Statistical Data of the Uranium Industry*.

[b]This definition is used in John Klemenic and Laurence Sanders, "Discovery Rates and Costs for Additions to Constant-Dollar Uranium Reserves," *Uranium Industry Seminar*, GJO-100(76) (Grand Junction, Colo.: USERDA, 1976), p. 276.

terms of a 1-year lag. In other words, we report the "following year's additions"; the amount of drilling done between January 1 and December 31 of a given year is compared with the net change in reserves plus cumulative production between January 1 and December 31 of the following year. We have chosen to calculate discovery rates on this basis for two reasons. First, "there is some lag [commonly ranging between 1 and 1½ years] in collecting and evaluating data from new discoveries and development activity."[27] And second, use of a 3-year moving average, or an aggregate discovery rate over a period of years, tends to create the misleading impression that discovery rates follow a smooth curve.

Given the data in figure 5–6 and table 5–8, we can make the following short-term projection: costs per foot drilled, in 1975 dollars, are likely to

Table 5–8
Discovery Rate for $15/lb U₃O₈, in Constant 1975 Dollars

Year	Following Year's Additions to $15/lb Reserves Plus Cumulative Production (10^3 tons U_3O_8)	Total Surface Drilling (10^6 ft)	Discovery Rate (lb U_3O_8/ft)	Cumulative Surface Drilling (10^6 ft)
1964	6.308	2.212	5.70	69.140
1965	9.184	2.113	8.69	71.253
1966	22.980	4.200	10.94	75.453
1967	28.660	10.764	5.33	86.217
1968	64.298	23.754	5.41	109.971
1969	69.643	29.855	4.67	139.826
1970	55.646	23.528	4.73	163.354
1971	22.745	15.452	2.94	178.806
1972	36.701	15.424	4.76	194.230
1973	24.794	16.421	3.02	210.651
1974	48.976	22.000	4.45	232.651

Source: Data on reserve additions and total drilling are from John Klemenic and Laurence Sanders, "Discovery Rates and Costs for Additions to Constant-Dollar Uranium Reserves," *Uranium Industry Seminar*, GJO-108(76) (Grand Junction, Colo.: USERDA, 1976), pp. 282–285. Data on cumulative drilling are from U.S. Energy Research and Development Administration, *Statistical Data of the Uranium Industry*, GJO-100(77) (Grand Junction, Colo.: USERDA, 1977), p. 75.

be around $4 per foot, and the discovery rate for $15 U₃O₈, in 1975 dollars, is likely to be around 4 pounds per foot, so costs of roughly $1 per pound U₃O₈ can be anticipated. For uranium in the $30 or $50 forward-cost categories, exploration costs will be lower.

In looking toward longer-term projections, we must observe that the discovery rate will eventually decline if an exhaustive search for conventional uranium resources is carried out. Using a different definition of the discovery rate, Lieberman has argued that it is already declining exponentially as cumulative drilling footage increases.[28]

The data in table 5–8 do not provide strong evidence of an exponential decline, but linear regression yields the following fit:

$$\frac{A}{D_e + D_d} = 10.02e^{-0.04808(D_e + D_d)} \qquad (5.31)$$

Integration under this curve, from zero to infinity, yields an estimate of 1,042,000 short tons U₃O₈. The Energy Research and Development Administration's estimate of probable resources, reserves, and cumulative production of $15 uranium, in 1975 dollars (as of January 1, 1975), was

1,370,100 short tons U$_3$O$_8$.[29] The discrepancy between these figures does not seem large enough, given the poor fit of equation 5.31, to suggest that it is imprudent to rely on official estimates of probable resources when estimating long-term uranium availability. However, if one accepts the hypothesis that "possible" and "speculative" resources can be discovered—an additional 850,000 short tons U$_3$O$_8$ in this case—one must assume that the discovery rate will not decline as rapidly as equation 5.31 indicates, unless access to uranium-bearing lands is severely restricted. We find that the data shown in figure 5–7 do not exhibit any trend with sufficient clarity to provide a basis for long-term projections.

Although it is not clear that discovery rates are declining, there can be no doubt that average ore grades are declining. The long-term trend in ore grades is shown in figure 5–8. We suggest that this trend cannot be explained simply by price increases and cost reductions; resource depletion must also play an important role.

Figure 5–9 shows a comparison of reserves plus cumulative production of $15/lb U$_3O_8$, in constant 1975 dollars, with cumulative surface drilling. A logistic curve fits the data quite well. These data are simply taken from table 5–8, given the fact that reserves plus cumulation production totaled 310,004 tons U$_3$O$_8$ at the end of 1965. The logistic curve has the following general form:

$$Y = \frac{c}{1 + aX^b} \qquad b < 0 \qquad\qquad (5.32)$$

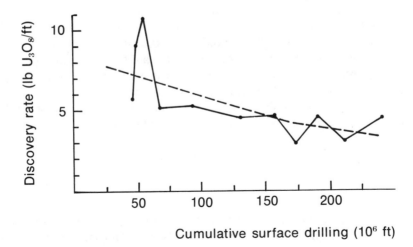

Figure 5–7. Discovery Rate for $15/lb U$_3O_8$, in Constant 1975 Dollars, Compared with Cumulative Surface Drilling.

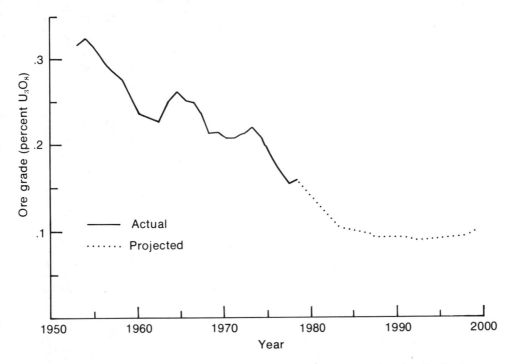

This graph is based on historical data provided in J. Fred Facer, Jr., "Uranium Production Trends," *Uranium Industry Seminar*, GJO-108(77) (Grand Junction, Colo.: USDOE, 1978), p. 177; and on a projection of future ore grades presented in John Klemenic and David Blanchfield, "Production Capability and Supply," at p. 218 of the same volume. The projection is based on an estimate of the maximum amount of uranium which could be produced from domestic resources at a forward cost no higher than $30 per pound U_3O_8.

Figure 5–8. Average Grade of Uranium Ore Processed in the United States, 1952–2000.

where Y = reserves plus cumulative production in thousand short tons
      U_3O_8
      X = cumulative surface drilling, in million feet

The discovery rate can be obtained simply by differentiating equation 5.32:

$$\frac{dY}{dX} = -\frac{abcX^{b-1}}{(1 + aX^b)^2} \qquad (5.33)$$

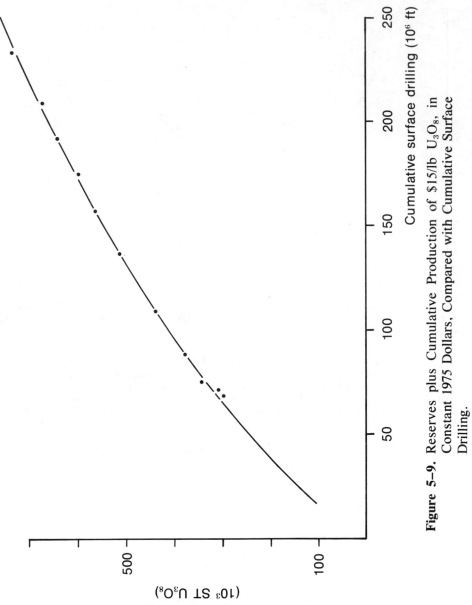

Figure 5–9. Reserves plus Cumulative Production of $15/lb U$_3O_8$, in Constant 1975 Dollars, Compared with Cumulative Surface Drilling.

The curve illustrated in figure 5–9 has the following form:

$$Y = \frac{2,050}{1 + 194.0X^{-0.8409}} \qquad (5.34)$$

We find that the discovery rate at 200 million feet of drilling is 3.67 lb/ft, but at 2 billion feet it is only 0.319 lb/ft, an order of magnitude lower. Reserves plus production at 2 billion feet would be 1.547 million short tons U_3O_8, only 75 percent of the ultimate resource total.

As figure 5–6 shows, drilling costs more than doubled between 1973 and 1976. The increase in oil prices in 1973 has caused a dramatic increase in the demand for drilling. The resulting shortage of drilling equipment and operators has caused a rise in drilling costs for oil. No doubt this explains part of the increase in costs for uranium. Since the future course of oil price regulation will have an important impact on oil drilling costs, it is very difficult to forecast costs per foot for uranium drilling.

Reclamation Cost

Until recently, it was common for uranium mines to be abandoned without any attempt at reclamation of the land. The Mining Law of 1872 "has no provision for protecting or rehabilitating lands covered by either mining claims or mineral patents," and consequently there are "scarred landscapes with potentially hazardous pits and unsightly waste material dumps."[30] As recently as 1973, the Atomic Energy Commission was rather sanguine about the situation:

> Little effort was made to control the environmental impact of uranium and vanadium mining in the Uravan Mineral Belt during the past 70 years or so. Nevertheless, the adverse effects have been relatively light. Very few of the mines encountered any appreciable water problems in mining. Consequently, discharge of water to nearby streams was not a serious problem. There is no problem of continued mine drainage after operations cease. The sandstone host rock in which the uranium is found is alkaline, containing from a few percent to 20 percent or more limestone. Thus, mine water is slightly alkaline, and the undesirable acid mine waters which frequently occur in coal mining from oxidation of sulfur, do not occur in uranium mining in this region.

> Outside of the disturbance of the surface in the course of exploration and mining, and the adverse effects of abandonment of some mines without proper protection and cleanup, there have been no permanent adverse effects from uranium-vanadium mining on the ecology of the Uravan Mineral Belt area that have been identified. On the other hand, surface disturbance has been widespread, and it affects scenic values.[31]

Since the AEC decided not to license uranium mines, their adverse effects on the ecology of an area could easily have escaped identification.

It is against this historical background that we must examine recent interest in surface disturbance. The ways in which current laws must be interpreted, and the cost of meeting those interpretations, are highly uncertain. For an estimate of current reclamation costs for open-pit mines, we may consider the case of the Bear Creek project, which involves an acquired lands lease in Medicine Bow National Forest. Reclamation costs of $1,390,350 were estimated by the company holding the lease (the Union Pacific Railroad), in connection with the mining of 3 million short tons of ore.[32] This excludes the cost of mill tailings stabilization, which will be discussed later. A reclamation cost of 46 cents per ton of ore is involved. A slightly lower figure can be derived from data prepared at Oak Ridge National Laboratory regarding a "model mine." An area of 1,050 acres would be disturbed by a mine producing 1,600 short tons of ore per day, 300 days per year, for 20 years.[33] At a cost of $3,000 per acre, reclamation would cost 33 cents per ton of ore. We have selected a figure of $3,000 per acre because it is the Department of the Interior's higher estimate of reclamation costs for the Sherwood uranium project.[34]

Substantially higher and lower estimates can be derived for particular mines. At the Sherwood project, "typically, the thickness of the mineralized section ranges from 30 to 100 feet, with mineralization concentrated in zones ranging from 1 foot to over 35 feet in continuous thickness."[35] Therefore, 7,950,000 tons of ore may be extracted from a mine disturbance area of only 155 acres.[36] At $3,000 per acre, reclamation will cost 6 cents per ton of ore. At the Bear Creek project, the most expensive mining and reclamation alternative considered—but not recommended—would involve an expenditure of $28,800,000 over and above the $1,390,350 just mentioned.[37] This would involve $9.60 per ton of ore. We consider such expenditures unlikely.

As a base case for our simulation model, we shall assume reclamation costs of 40 cents per ton of ore from open-pit production. A major research project would be required to establish a set of cost estimates for particular site conditions and legal requirements, and we are unprepared to attempt such an investigation.

For underground mines, evidence suggests that reclamation costs are so low as to be dwarfed by the uncertainties associated with other types of costs. On the Navajo-Exxon project, a 1,500 ton per day mine is estimated to require "about 10 acres for the necessary surface headworks" and 5 acres for mine waste.[38] On the Ute Mobil project, a 500 ton per day mine is estimated to "require a clearing of approximately five acres of surface area."[39] By contrast, open-pit production at the Bear Creek project involves the disturbance of 1,786 acres for a 1,000 ton per day mine.[40] We shall assume reclamation costs for underground production to be negligible.

Tailings Stabilization Cost

The release of radon gas from uranium mill tailings presents a substantial long-term radiation hazard.[41] The legal requirements for management of these tailings will largely be determined by the findings of a generic environmental impact statement on uranium milling which is currently being prepared by the Nuclear Regulatory Commission.[42] However, estimates of the lower and upper limits of tailings stabilization costs can be made on the basis of data published in two of the Commission's recent environmental impact statements.

The minimum requirement is simply to cover the tailings with soil and/or clay. For the Bear Creek project, an expenditure of $674,450 is associated with the milling of 3 million tons of ore; thus a cost of 22 cents per ton of ore is involved.[43] At the Lucky Mc uranium mill 5,500,000 tons of tailings have already been generated, and operation at 1,200 tons per day, 300 days per year, for 5.9 years will generate an additional 2,124,000 tons. An expenditure of $1,800,000 is equivalent to 24 cents per ton of ore.[44]

The maximum requirement is to fix the tailings in concrete after removing them to a mine pit. For the Bear Creek project, an additional $39,000,000 is required, raising the tailings stabilization cost to $13.00 per short ton of ore.[45] For the Lucky Mc uranium mill, $48,100,000 is required, or $6.25 per ton of ore.[46]

The radiation hazard from uranium mill tailings is principally related to the amount of uranium present in the ore delivered to the mill, not to the volume of this ore. It is plausible, therefore, that different standards will be established for different grades of ore. Mill feed is commonly blended to achieve a relatively constant average grade of ore, so it seems unlikely that low-grade ores would be milled separately to take advantage of lower tailings stabilization costs. However, the establishment of strict requirements, such as the fixation of tailings in concrete, would have an important effect on the cutoff grade of ore.

In the development of our computer simulation model, we shall treat both reclamation costs and tailings stabilization costs as additions to operating costs per ton of ore regardless of ore grade. In this way, the effect on cutoff grades will be incorporated. We shall assume minimum tailings stabilization costs, that is, 25 cents per ton of ore, and then repeat the calculations for costs of $6.00 per ton of ore.

Notes

1. Battelle–Pacific Northwest Laboratories, *Assessment of Uranium and Thorium Resources in the United States and the Effect of Policy Alternatives*, PB 238 658 (Springfield, Va.: NTIS, 1974), pp. 6.21–6.23.

2. John Klemenic, *An Estimate of the Economics of Uranium Concentrate Production from Low Grade Sources* (Grand Junction, Colo.: USAEC, 1974), pp. 7–10, 20–23. See also U.S. Atomic Energy Commission, *Nuclear Fuel Resource Evaluation: Concepts, Uses, Limitations*, GJO-105 (Grand Junction, Colo.: USAEC, 1973), pp. 7–14.

3. U.S. Energy Research and Development Administration, *Statistical Data of the Uranium Industry*, GJO-100(77) (Grand Junction, Colo.: USERDA, 1977), p. 106.

4. We refer to studies by the Grand Junction Office.

5. Most of the notation used in this chapter is consistent with that used in Wallace E. Tyner and Robert J. Kalter, *A Simulation Model for Resource Policy Evaluation*, Cornell Agricultural Economics Staff Paper No. 77-28 (Ithaca, N.Y.: Department of Agricultural Economics, Cornell University, 1977). However, some differences exist; for example, F is used to describe production profiles, but we define it differently from the way Tyner and Kalter define it.

6. John F. Hogerton, Clifton H. Barnes, William A. Franks, and Robert J. McWhorter, *Report on Uranium Supply: Task III of EEI Nuclear Fuels Supply Study Program*, (New York: S.M. Stoller, 1975), p. 92.

7. John Klemenic, *Examples of Overall Economics in a Future Cycle of Uranium Concentrate Production for Assumed Open Pit and Underground Mining Operations* (Grand Junction, Colo.: USAEC, 1972), pp. 1, 7.

8. These equations describe curves which fit the data points extremely well. Since the data are only estimates of costs for the "model" mine-mill complex and not empirical observations, the precise fit does not really prove anything, and we do not present statistics to describe it.

9. Costs for mine plant and mine equipment were combined in one category in the published source, so we shall assume a one-third/two-thirds split of this category.

10. U.S. Energy Research and Development Administration, *Statistical Data of the Uranium Industry*, GJO-100(77), p. 59.

11. Klemenic, "An Estimate of the Economics," pp. 7–10, 20–23.

12. Klemenic, "Examples of Overall Economics," pp. 2, 7; Klemenic, "An Estimate of the Economics," pp. 7, 8, 10, 20, 21, 23.

13. See Arthur D. Little, Inc., *Competition in the Nuclear Power Supply Industry*, TID UC-2 (Cambridge, Mass.: Arthur D. Little, Inc., 1968), pp. 163–164.

14. See "Commonwealth Edison's Purchase of Cotter Sets Precedent," *Nuclear Industry*, May 1970, p. 32.

15. John A. Graves. "Open Pit Uranium Mining," *Mining Engineering* 26,8 (August 19, 1974).

16. Klemenic, "Examples of Overall Economics," pp. 8, 14.

17. A 95 percent recovery rate can be derived from the figures in Klemenic, "Examples of Overall Economics," pp. 2, 7. We used this assumption for the calculation of costs per short ton of ore.

18. Klemenic, "An Estimate of the Economics," pp. 7–10, 20–23.

19. U.S. Department of the Interior, Bureau of Indian Affairs, *Final Environmental Statement of the Approval by the Department of the Interior of a Lease of the Ute Mountain Ute Tribal Lands for Uranium Exploration and Possible Mining*, FES 75-94 (Albuquerque, N.M.: USDI, 1975), p. 31.

20. Klemenic, "An Estimate of the Economics," p. 4.

21. U.S. Energy Research and Development Administration, *Statistical Data of the Uranium Industry*, GJO-100(77), p. 106.

22. Ibid., pp. 34, 36, 38.

23. An ore grade distribution for a hypothetical deposit is illustrated in U.S. Atomic Energy Commission, *Nuclear Fuel Resource Evaluation*, GJO-105, p. 13.

24. Tyner and Kalter, *A Simulation Model for Resource Policy Evaluation*, p. 19.

25. Klemenic, "Examples of Overall Economics," pp. 30–36.

26. John Klemenic and Laurence Sanders, "Discovery Rates and Costs for Additions to Constant-Dollar Uranium Reserves," *Uranium Industry Seminar*, GJO-100(76) (Grand Junction, Colo.: USERDA, 1976), pp. 271–294.

27. U.S. Energy Research and Development Administration, *Statistical Data of the Uranium Industry*, GJO-100(76) (Grand Junction, Colo.: USERDA, 1976), p. 16.

28. M.A. Lieberman, "United States Uranium Resources—An Analysis of Historical Data," *Science* 192, 4238 (30 April 76):431–436.

29. U.S. Energy Research and Development Administration, *Statistical Data of the Uranium Industry*, GJO-100(77), pp. 24, 42.

30. U.S. General Accounting Office, *Modernization of 1872 Mining Law Needed to Encourage Domestic Mineral Production, Protect the Environment, and Improve Public Land Management*, B-118678 (Washington: USGAO, May 1975), p. 24.

31. U.S. Atomic Energy Commission, *Leasing of AEC Controlled Uranium Bearing Lands, Colorado, Utah, New Mexico*, WASH-1523 (Washington: USAEC, 1972), pp. 16–17.

32. U.S. Nuclear Regulatory Commission, U.S. Department of the

Interior, and U.S. Department of Agriculture, *Final Environmental Statement Related to Operation of Bear Creek Project, Rocky Mountain Energy Company*, NUREG-0129 (Washington: USNRC, 1977), pp. J-3, 3-1.

33. W.R. Grimes, *Natural Resource Use in and Effluents from Mining and Milling Operations for Various Reactors* (Oak Ridge, Tenn.: Oak Ridge National Laboratory, October 1976), pp. 4, A-19.

34. "Estimating the per acre cost at this time, based upon another similar operation, would produce a figure of $2,000 to $3,000 per acre." U.S. Department of the Interior, Bureau of Indian Affairs, *Final Environmental Impact Statement, Sherwood Uranium Project, Spokane Indian Reservation*, FES 76/45 (Portland, Ore.: USDI, 1976), p. 4-43.

35. Ibid., p. 2-39.

36. Ibid., pp. 1-2, 4-26.

37. U.S. Nuclear Regulatory Commission et al., *Final Environmental Statement Related to Operation of Bear Creek Project*, p. 10-7.

38. U.S. Department of the Interior, Bureau of Indian Affairs, *Final Environmental Impact Statement, Navajo-Exxon Uranium Development*, FES 76-60 (Billings, Montana: USDI, 1976), pp. 1-10, 1-12, 1-34.

39. U.S. Department of the Interior, Bureau of Indian Affairs, *Final Environmental Statement of the Approval by the Department of the Interior of a Lease of the Ute Mountain Ute Tribal Lands for Uranium Exploration and Possible Mining*, FES 75-94 (Albuquerque, N.M.: USDI, 1975), pp. 10, 12.

40. U.S. Nuclear Regulatory Commission et al., *Final Environmental Statement Related to Operation of Bear Creek Project*, pp. 1-1, 10-11.

41. Robert O. Pohl, "Health Effects of Randon-222 from Uranium Mining," *Search* 7,8 (August 1976):345–354.

42. 41 F.R. 22430 (June 3, 1976); and 42 F.R. 13874 (March 14, 1977).

43. U.S. Nuclear Regulatory Commission et al., *Final Environmental Statement Related to Operation of Bear Creek Project*, pp. 3-1, J-3.

44. U.S. Nuclear Regulatory Commission, *Draft Environmental Statement Related to Operation of Lucky Mc Uranium Mill, Utah International, Inc.*, NUREG-0295 (Washington: USNRC, 1977), pp. 3-12, 3-16.

45. U.S. Nuclear Regulatory Commission et al., *Final Environmental Statement Related to Operation of Bear Creek Project*, pp. 10-13 to 10-18.

46. U.S. Nuclear Regulatory Commission, *Draft Environmental Statement Related to Operation of Lucky Mc Uranium Mill*, pp. 10-3 to 10-4.

6 The Domestic Uranium Resource Base

We shall begin our discussion of the resource base by defining it, and by distinguishing three categories of resources. Next, we identify the relationship of resource estimates to the data requirements of our study and to the public policy issues affecting nuclear power. With this introduction, we proceed to an examination of two of the methods of resource estimation: geological analogy and discovery-rate models. Finally, we review the extent to which current knowledge can provide answers to some of the public-policy questions identified earlier.

Definitions

By *domestic uranium resource base* we mean the total amount of uranium in the earth's crust underlying the United States. In other words, we begin with the broadest definition possible. There are many different ways in which one might distinguish categories of resources; we prefer the conceptual framework used by the Grand Junction Office of the Department of Energy (formerly part of the Energy Research and Development Administration and originally part of the Atomic Energy Commission). This framework was described in 1973. Three major categories exist: reserves, potential resources, and unassessed resources.

> A *uranium ore reserve* is defined by the AEC as an estimate of the quantity of uranium in known deposits which can be produced at or below a stated cost per pound termed the "cutoff cost."

> . . . *Potential resources* of uranium are those surmised to occur in unexplored extension of known deposits, or in undiscovered deposits within or adjacent to uranium areas, or in other favorable areas. They are expected to be discoverable and exploitable at a given cost. A potential estimate is a judgment of the amount of undiscovered uranium in areas about which enough is known of the geology to indicate the nature and extent of the environment favorable for the occurrence of uranium [italics added].[1]

Unassessed resources include everything outside reserves and potential.

Reserves

The Department of Energy does not publish information on different categories of reserves. However, the American Institute of Mining Engineers, the American Institute of Professional Geologists, and the Society of Economic Geologists have proposed a general classification of mineral reserves according to the risk faced by companies interested in developing them:

> Proven Reserves are ore reserves so extensively sampled that the tonnage, grade, geometry, and recoverability of the ore within the block or blocks of ground under consideration can be computed with sufficient accuracy so that the uncertainties involved would not be a factor in determining the positive feasibility of a mining operation.
>
> Probable Reserves are ore reserves for which sufficient continuity of dimensions and grade can be assumed for preliminary financial planning, but for which the risk of failure in continuity is greater than for proven ore.
>
> Possible Reserves are mineralized material of which the dimension and grade are based on geologic correlation between samples so widely spaced or so erratic that additional exploration is required to establish whether ore reserves are present.[2]

Under this set of definitions, "possible reserves" include a portion of what the Department of Energy calls "potential resources."

Potential Resources

Potential resources include three categories, as follows:

> "Probable" potential resources are those estimated to occur in known productive uranium districts (where "productive" means that past production plus known reserves exceeds 10 tons U_3O_8):
>
> 1. In extensions of known deposits, or
> 2. In undiscovered deposits within known geologic trends or areas of mineralization.
>
> "Possible" potential resources are those estimated to occur in undiscovered or partly defined deposits in formations or geologic settings productive elsewhere within the same geologic province.
>
> "Speculative" potential resources are those estimated to occur in undiscovered or partly defined deposits:
>
> 1. In formations or geologic settings not previously productive within a productive geologic province, or
> 2. Within a geologic province not previously productive.[3]

Here the distinguishing criterion is the geological similarity between the area with potential resources and some area for which past production plus known reserves exceeds 10 tons U_3O_8.

In general, the risk of failure to discover reserves is likely to be lower in extensions of known deposits or trends than in areas which are merely in productive formations or geologic settings; likewise, the latter are generally less risky than areas which are merely in productive geologic provinces. However, geological similarity is not synonymous with low financial risk. Known deposits or trends are precisely those which have already been explored, and wherever the exploration has been thorough, there is only a slim chance that major deposits have somehow escaped everyone's attention. The viewpoint of the Getty Oil Corporation regarding the Shirley Basin of Wyoming in 1975 illustrates this point:

> We believe we have completely drilled out the ore deposits in this area and were thus surprised to learn that the AEC has assigned an additional 10% ± potential ore to this area. We believe the unusually extensive and dense drilling precludes the possibility of additional ore.[4]

Given this situation, the labels *probable, possible,* and *speculative* are somewhat misleading. For any particular area there is only one estimate of potential resources; the Department of Energy does not estimate what is probably there, what is possibly there, or what is speculatively there. It may be argued that if a probable resource lies in a relatively unexplored area, it is probably there; but if it lies in a thoroughly explored area, its presence is speculative.

Unassessed Resources

Unassessed resources are not broken down into official categories at all. However, we would like to propose these categories, if only to clarify the definition of unassessed resources:

1. *Economically identifiable unassessed resources* are those which are located in areas for which no attempt has been made to estimate potential resources, but if such an attempt were made, potential would be found. In other words, both the environments favorable for the occurrence of uranium and the cost of extracting it could be identified. We expect that some of the uranium in Alaska falls under this category.

2. *Geologically identifiable unassessed resources* are those for which environments favorable for the occurrence of uranium could be, or have been, identified, but the cost of discovering and exploiting the deposits cannot be estimated with enough accuracy or confidence to place the resources in standard cost categories. (There are now four such

categories—$10, $15, $30, and $50—based on forward costs per pound of U_3O_8 recovered.)

3. *Unidentifiable unassessed resources* are those which could not be identified on the basis of present geological theory, except by close-spaced drilling. Because our understanding of the formation of uranium deposits is incomplete, these deposits cannot be found except by drilling in areas where no one would expect uranium. Since this kind of drilling is a poor investment, most of these deposits may never be discovered.

Mineral Inventories

It should be emphasized that all reserves and potential resources are defined in terms of cost categories. An alternative approach to resource assessment is to establish some geological parameter, or parameters, in lieu of cost. For example, ore-grade categories can be used. When known uranium deposits are involved, a resource estimate based on ore grades is called a *mineral inventory*. The degree of certainty with which this inventory is known is roughly comparable to the certainty with which reserves are known. The Energy Research and Development Administration first published statistics on uranium mineral inventories in 1977.[5]

Usefulness of Potential Resource Estimates

There can be no question that information on reserves is necessary to justify major investments in mine and mill production. However, estimates of potential resources do not have any immediate practical application. It is important to distinguish judgments about where to look for uranium from estimates of how much is likely to be discovered. A geologist may compare various geological settings, and rank them in order of favorability of occurrence for uranium, without attempting to state whether or not any of them contain uranium. He may feel prepared to answer the question "If uranium is out there, where is it most likely to be?" but not the question "How much uranium is out there?" Answers to the latter question are not necessary to guide the day-to-day operations of an exploration program. This fact is reflected in industry comments on the Department of Energy's resource evaluation program.[6]

It is reasonable to ask, therefore, why serious attention should be given to potential resource estimates. There are two reasons. First, the topic is one of the considerations involved in estimates of uranium prices, and second, it is related to questions of public policy which are likely to receive at least as much attention as the disposal of uranium resources on federal lands.

Implications for Uranium Prices

Since uranium ore is not a renewable resource, the effect of depletion is to raise costs. A simple theoretical relationship between drilling and non-drilling expenditures, showing the effect of depletion, is shown in figure 6–1. Let us assume that the demand for U_3O_8 is constant between two

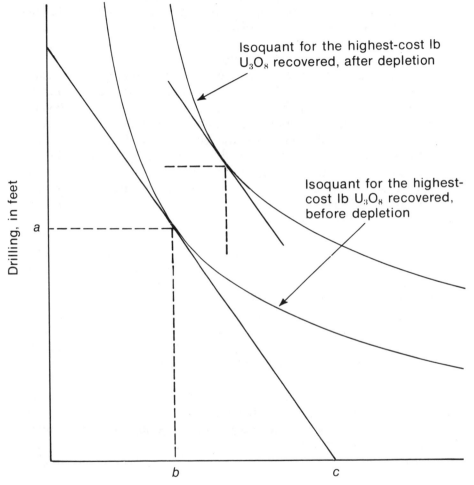

Present value of nondrilling expenditures, in dollars

Note: Cost per foot of drilling, in present dollars $= (c - b)/a$; price per pound of U_3O_8, in present dollars $= c$; discovery rate for highest-cost lb $U_3O_8 = 1/a$.

Figure 6–1. Theoretical Relationship Between Drilling and Nondrilling Expenditures.

years (if we assume inelastic demand, this means annual production is constant). Furthermore, let us assume a perfectly competitive market in equilibrium with a uniform price paid by all purchasers. In a given year, the cost of the highest-cost pound of U_3O_8 will equal the price. The effect of depletion is to shift the isoquant for this highest-cost pound outward. If the cost per foot of drilling is constant (as shown) or increasing, in present-valued dollars, then depletion will result in a higher marginal cost of U_3O_8 and therefore a higher price. The supply curve will be shifted upward. However, the discovery rate may or may not decrease depending on the way in which the isoquant shifts. In figure 6–1, the discovery rate falls; we may consider this a likely situation.

If the cost per foot of drilling decreases, the depletion may not result in a higher marginal cost of U_3O_8. However, the discovery rate is virtually certain to decrease; only a rather drastic change in isoquant could prevent this effect. If the efficiency of mining and milling operations is increased, or if wages, capital-equipment costs, fuel costs, and/or leaching-solution costs decrease, depletion may not result in a shift of the isoquant. These other factors could outweigh depletion and shift the isoquant inward, that is, shift the supply curve downward. It should be clear from this discussion that the effects of resource depletion cannot always be observed in higher prices and lower discovery rates, even when demand is assumed to hold constant. Given changes in demand, the whole picture is so complicated that it would be very difficult to construct a reliable empirical measure of the effects of depletion on the highest-cost increment of production. For total production, the picture is even more indeterminate.

The usefulness of potential resource estimates, in connection with our simulation model, is their contribution to the realm of information needed to provide an overall picture of long-term trends in uranium prices. The task of selecting possible price scenarios to be evaluated is one which involves a great deal of subjective judgment. The domestic resource base is just one of many factors to be considered.

It would be desirable, of course, to have potential resource estimates associated with specific tracts of federal land, vacant or claimed. Such information could be used to assess the potential impact of a uranium leasing program on the uranium market, given different assumptions about lease schedules. Unfortunately, no data of this type are available.

Public-Policy Issues

There are several public-policy questions related to the domestic resource base. Controversy over these issues may be just as intense as controversy over reform of the Mining Law of 1872, and although they are not central to our research, we feel that some recognition of their importance is due. Four questions stand out.

1. Does the domestic resource base permit the lifetime fuel requirements of the nuclear capacity envisioned by the Carter administration's National Energy Plan to be met without dependence on imported uranium, and without excessive environmental cost?
2. Is the domestic resource base sufficiently limited that import restrictions may be required to prevent low-cost foreign uranium from taking a large market share?
3. How much money must the federal government spend on uranium resource assessment, if the nuclear capacity goals of the National Energy Plan are to be achieved?
4. How much money should the federal government spend on breeder reactor development?

We shall return to the first question after some of the methodology of potential resource estimation has been discussed.

The public-policy issues just listed require more information about the domestic resource base than private industry would find necessary or worthwhile to obtain. The only portion of the resource base which is known with reasonable certainty is the reserve category. In order to resolve some of the preceding questions, it may be appropriate to provide incentives, or requirements, for industry to expand its reserve holdings beyond the level which would be dictated by market forces. Alternatively, it may be considered desirable for the government to locate reserves, with or without participation by private contractors. Unfortunately, these actions are rarely considered in the published literature on potential resources. It is important to note that potential resource estimates are simply based on speculative judgments. There is no way that improvements in the procedures used to make these estimates will ever yield data which are as reliable and accurate as reserve data.

Methods of Resource Estimation

$30 Resources

The method used by the Department of Energy to estimate potential resources at forward costs up to $30 per pound U_3O_8 is that of geological analogy:

Unexplored areas are compared with "control" areas in which the uranium production and ore reserves are known. . . . The following formula summarizes the basic approach to estimating potential.

$$P = T \times N \times F$$

where P = potential in area being evaluated.

$\quad\quad T$ = mineralization density of the control area; tons U_3O_8 per square mile (or other unit) of explored favorable host rock.

$\quad\quad N$ = number of square miles (or other unit) of unexplored favorable ground in the area being evaluated.

$\quad\quad F$ = geological favorability factor: the relative favorability of the area being evaluated compared to the control area—based on geology, the results of exploration, presence of uranium mineralization, and intensity and extent of anomalous radioactivity.[7]

Unfortunately, the state of the art of uranium geology does not provide a basis for establishing favorability factors with much confidence. According to a representative of Kerr-McGee Corporation, the nation's largest uranium mining and milling firm, there is a lot to be learned. "We in no case know the source of the uranium, although in most cases multiple sources seem available. We in most cases do not know how the uranium has migrated to the sites of deposition; and in most cases we cannot say why a uranium deposit has formed at one place and not another."[8] Similar remarks were included in a study prepared by the S.M. Stoller Corporation.[9]

The dominant philosophy behind the Department of Energy's potential resource estimates is that uranium is most likely to be found in the same formations where it has been found before. We have noted that the labels *probable, possible,* and *speculative* have been assigned to geological areas in a way that supports this philosophy. Further evidence of such support exists in the fact that six formations account for the major share of both past production and probable resources.

Roughly 90% of past sandstone production has come from three geologic formations: the Morrison Formation and late Jurassic age; the Wind River Formation of early Tertiary age; and the Chinle Formation of Triassic age. In the present reserve situation, the Morrison and Wind River Formations have retained their importance, while the importance of the Chinle Formation has greatly decreased.[10]

The Battle Spring, Wasatch, and Fort Union Formations are all Eocene or Paleocene sandstones associated with the Wind River Formation. These, plus the Morrison and Chinle Formations, represent roughly 70 percent of probable resources, as table 6–1 shows.

Table 6–1
Probable Uranium Resources in the United States, as of January 1, 1976

Symbol Used by the Department of Energy	Region	Host Rock	$30 Probable Resources ($10^3$ tons U_3O_8)	Date of Uranium Discovery
B4, B5	Wyoming Basins	Battle Spring, Wasatch, Wind River, and Fort Union Formation[a]	288.1	1951
F3	Great Plains	Fort Union Formation	9.6	1951
A12–A16	Colorado Plateau	Morrison Formation	355.2	1880
E13	Colorado and Southern Rockies	Morrison Formation	5.7	1880
A20–A23	Colorado Plateau	Chinle Formation	47.2	1948
		Total	705.8	
		All other locations	354.2	
		Total U.S.	1060.0	

Source: U.S. Energy Research and Development Administration, *National Uranium Resource Evaluation, Preliminary Report*, GJO-111(76) (Grand Junction, Colo.: USERDA, 1976), pp. 5, 41, 49, 68, 74; R.J. McWhorter, D.R. Hill, E.A. Noble, J.F. Hogerton, and C.H. Barnes, *Uranium Exploration Activities in the United States*, EPRI EA-401 (Palo Alto, Calif.: Electric Power Research Institute, 1977), pp. B-33, B-34.
[a]These are all Eocene or Paleocene sandstones.

The Atomic Energy Commission estimated potential resources at forward costs up to $30 per pound to be 1.6 million short tons U_3O_8, as of yearend 1969. This figure did not change until the definition of *potential* was changed from "estimated additional resources" (one category) to "probable," "possible," and "speculative." As of January 1, 1975, the probable resources were estimated to be 1.14 million tons U_3O_8; two years later, this figure fell to 1.09 million tons U_3O_8.[11] This record shows that no great changes in the potential resource outlook have occurred since 1969, as far as $30 resources are concerned.

Higher-Cost Resources

A very different picture emerges when one examines the $50 and $100 categories. The possibility of commercial interest in measurement of these resources has arisen only recently as a result of uranium price increases. In 1969 it was estimated that 8 million tons U_3O_8 were available from "reasonably assured" and "estimated additional" resources in the $30 to $50 increment; by 1977 the comparable figure had dropped to 0.44 million tons U_3O_8, based on reserves and probable resources. The outlook was summarized in a speech by Robert Nininger, Assistant Director for Raw Materials, in September 1975. At that time Nininger doubted that uranium resources costing substantially more than $30 per pound could be relied upon for future supplies because of the environmental problems involved in mining commercial quantities of low-grade ore.[12]

In retrospect, the Atomic Energy Commission's assurance of "reasonably assured" resources at forward costs above $30 per pound U_3O_8 does not appear very reasonable. If our sandstone resources are exhausted, the "next best" major source of uranium is Chattanooga shale, for which the dollar costs are unknown and the environmental effects are likely to be politically unacceptable. Although it is very difficult to estimate political acceptability, we may surmise that the outlook for regulatory approval becomes very unfavorable when a ton of uranium ore from a proposed mine will yield less electrical energy than a ton of coal. For a new power plant burning Eastern coal, we may estimate the yield very simply:

$$\frac{h_c}{H} = \frac{2.4 \cdot 10^7 \text{ Btu/ST coal}}{9.3 \cdot 10^3 \text{ Btu/kWeh}} = 2581 \text{ kWeh/ST coal} \qquad (6.1)$$

where h_c = heat content of coal, in Btu/ST coal
H = heat rate of power plant, in Btu/kWeh

For a nuclear plant utilizing uranium mined from sandstone ores, we may estimate the recovery rate of U_3O_8 as a function of ore grade, according to equation 5.12. At an ore grade of 157 parts per million (ppm), or 0.0157 percent U_3O_8, the energy yield is comparable to that of coal:

$$GxD = 157 \cdot 10^{-6} \text{ ST } U_3O_8/\text{ST ore}$$
$$\cdot 0.6578 \cdot 2.5 \cdot 10^7 \text{ kWeh/ST } U_3O_8 \qquad (6.2)$$
$$= 2582 \text{ kWeh/ST ore}$$

where G = ore grade, in ST U_3O_8/ST ore
  x = recovery rate
  D = duty factor, in kWeh/ST U_3O_8

The average grade of Chattanooga shale is 50 to 70 ppm uranium (not U_3O_8). Even if it could be milled with a recovery rate above 66 percent, it could not possibly meet this criterion of political acceptability.[13] Using a comparison based on Western coal, a high recovery rate, and a high duty factor, one might present the shale in a more favorable light, but we believe that such a presentation would not accurately reflect the environmental problems.

One indication of the political obstacles to mining low-grade ores is the recent Swedith experience. Ten thousand demonstrators appeared at a mine at Billingen to protest the extraction of shales containing 250 to 325 ppm of uranium. The government-owned mining company, LKAB, decided to defer the start of mining and reduce the scale of the operation from 6 million metric tons of shale per year to 1 million, despite the fact that Sweden had no other significant uranium resources.[14]

It has been estimated that the overburden moved per ton of Chattanooga shale would be roughly equal to that moved per ton of Eastern coal.[15] Since this shale has never been mined, and since the management of mill tailings could represent a major earth-moving effort, this estimate must be regarded as quite speculative.

As of January 1, 1977, uranium-bearing properties with an average grade of ore below 300 ppm represented less than 1 percent of the U_3O_8 contained in $30 reserves plus cumulative production.[16] In view of the Swedish controversy it is possible that regulations prohibiting the mining of ores assaying 300 ppm, or some other grade, will be established. Such restrictions may be extended from limitations on the average grade of a deposit to limitations on the cutoff grade for mining operations at all deposits. We do not incorporate these sorts of regulatory constraints in the simulation model because no attempt has been made to establish them. However, the issue must be seriously examined when higher-cost reserves are being considered for domestic production.

Discovery-Rate Models

If the discovery rate, in pounds U_3O_8 per foot of drilling, is graphed as a function of cumulative drilling footage, the integral under the curve is obviously equal to reserves plus cumulative production. If one could project future discovery rates out to the drilling footage at which the rate approaches zero, then one could estimate the available resource base. We may define the integral under this projection to be "recoverable resources," which can be added to the integral of past discovery rates, that is, reserves plus production. An estimate of recoverable resources is more easily applied to the policy questions we have listed than an estimate of potential resources, since the definition of the latter does not include any consideration of which areas are open to exploration. However, a reliable estimate of recoverable resources is very difficult to make, because it involves more economic factors than an estimate of potential resources.

M.A. Lieberman has stated that a projection of future discovery rates can be made by fitting an exponential curve to observed data and extrapolating this curve to infinite drilling footages. The exponential form is chosen simply because it "fits" the data, as Lieberman has used them.[17] However, as we have seen in chapter 5, the trend in additions to constant-dollar reserves plus cumulative production per foot of total surface drilling does not fit any exponential curve very well. Moreover, as Harris has observed, "There are no theoretical grounds for the negative exponential as the appropriate model."[18] Some of the theoretical issues involved in projecting discovery rates have already been discussed, and it is clear that the economic picture is complicated. By developing a theoretical model which takes into consideration the cost per foot of drilling and price expectations for U_3O_8, one might attempt to explain past discovery rates for marginal (high-cost) ores. An explanation of discovery rates for all ores would be more speculative.

Major Discoveries

The estimation of potential resources is complicated by the role of major discoveries, of which three types can be distinguished.

1. Discovery of a large deposit. Since economies of scale in uranium mining and milling are important, large deposits make up a greater proportion of reserves than of mineral inventories. The distribution of reserve deposits of more than 25,000 short tons of ore is shown in figure 6–2; the number of deposits appears to decrease linearly in proportion to the logarithm of deposit size. Data on the distribution of mineral inven-

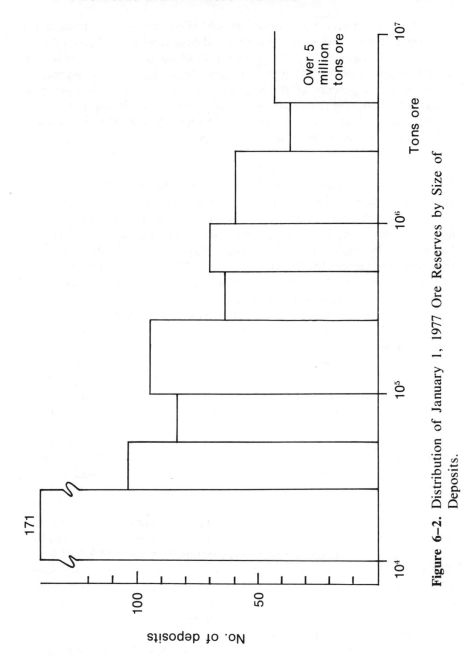

Figure 6–2. Distribution of January 1, 1977 Ore Reserves by Size of Deposits.

tories by deposit size have not been published, but we might expect a lognormal distribution, as in the case of oil fields and natural gas fields.[19]

Forty-three deposits contain 59 percent of the $30 reserves recorded as of January 1, 1977.[20] If we assume that the average size of deposits is the same in 59 percent of the $30 probable resources, then sixty-nine deposits will contain that share of probable resources. If these resources are discovered over a 23-year period, 1978 through 2000, the largest three deposits discovered in any year should represent over half of that year's discoveries, *on average*. Of course, a large deposit is unlikely to be entirely converted from probable resources to reserves in one year. Nevertheless, the industrywide discovery rate can fluctuate from year to year in response to the randomness of major discoveries. Although the geological favorability of a particular formation or area may be overestimated if a major find has already been made and underestimated if a major find is yet to be made, these errors should "average out" in the estimation of potential resources for the whole country. Therefore, the concentration of reserves in large deposits presents a disadvantage for discovery-rate models relative to geological analogies.

2. Discovery of a major uranium-bearing district yielding sandstone deposits. When a new district is found, one discovery leads to another, and the industrywide discovery rate will rise rapidly. With the saturation of known districts, the discovery rate falls. Consequently, the estimate of recoverable resources yielded by a discovery-rate model will be influenced by a small number of random events. This is particularly true when the industry's geological understanding of uranium deposition is insufficient to indicate where to look for new districts.[21]

3. Discovery of a major nonsandstone deposit. It is well known that most of the world's uranium reserves do not occur in sandstone deposits.[22] Since 1970, some exploration efforts have been devoted to nonsandstone deposits; 16.7 percent of total exploration costs in 1976 were devoted to them.[23] A very thick volcanic deposit was found in Utah in 1977,[24] but on the whole, results have been disappointing. "Sought, but without reported successes, were calcrete-type deposits and vein-like deposits of Australia, and porphyry-type deposits of South West Africa as well as ancient uraniferous conglomerates of the Canadian and South Africa type."[25] If a breakthrough occurs in finding nonsandstone deposits, discovery rates should rise and estimates of both recoverable resources and probable resources should jump upward.

There are other methods of resource estimation, and they have been reviewed elsewhere.[26] However, we believe that the Department of Energy's procedure for estimating $30 potential resources is presently the most reliable basis for assessing the domestic resource base. A discussion of alternative procedures would not provide an essential contribution to our evaluation of policies regarding federal lands.

Coverage of Reactor Lifetime Requirements

The leading public-policy issue cited earlier was the question of whether the domestic resource base permits the lifetime fuel requirements of the nuclear capacity envisioned by the Carter administration's National Energy Plan to be met without dependence on imported uranium and without excessive environmental cost. There are several ways of measuring the adequacy of the resource base relative to uranium requirements, but we prefer this formulation. Prior to the National Energy Plan, the question could be stated with reference to official nuclear capacity projections, such as those issued by the Atomic Energy Commission and the Energy Research and Development Administration.

There are four dominant factors which tend to determine the overall outlook for coverage of reactor lifetime requirements:

1. The date for which installed capacity is to be covered.
2. The projection of installed nuclear capacity.
3. The commercial availability of Chattanooga shale and other low-grade shales.
4. The projection of duty factor, averaged over each reactor's lifetime.

Over the past 8 years, different authors have used a variety of assumptions regarding these four topics and thereby arrived at different assessments of uranium availability. Since this debate has influenced price expectations, we shall briefly review its evolution.

Technological Optimism

In September 1971 the director of the Atomic Energy Commission's Division of Raw Materials presented a paper at the fourth United Nations International Conference on the Peaceful Uses of Atomic Energy. The basic point of the paper was to assure the rest of the world that the United States could increase its domestic nuclear capacity very rapidly without running out of uranium. A forecast of 1,500 GWe (gigawatts electrical) by the year 2000 was presented.[27] If this had been entirely composed of light-water reactor capacity, lifetime requirements of 5 short tons U_3O_8 (in concentrate) per megawatt (electrical) would have been projected.[28] Thus requirements of 7.5 million tons U_3O_8 could have been compared with "reasonably assured" and "estimated additional" resources (including byproduct) of 2.41 million tons U_3O_8 (in ore) at $30 per pound as of January 1971.[29] Such an outlook would not have appeared very promising.

One might ask, therefore, how it could be possible for the Atomic Energy Commission to assure the rest of the world that the United States would not face a shortage of uranium. The answer is simple: lifetime requirements were not considered at all, and duty factors were assumed to be very high. Uranium resources were compared with delivery requirements through the year 2000. For 1,500 GWe, the cumulative requirement (1970–2000) was 1.7 to 1.8 million tons U_3O_8, or 1.13 tons per megawatt installed in 2000.[30] By contrast, the Department of Energy forecast under the National Energy Plan involved a cumulative requirement (1977–2000) of 1.06 million tons U_3O_8 for 380 gigawatts, or 2.79 tons per megawatt installed in 2000 (assuming a 0.25 percent tails assay).[31] The 1971 forecast assumed a rapid introduction of high-temperature gas reactors and liquid-metal fast-breeder reactors. Cumulative requirements were compared with reasonably assured and estimated additional resources (including byproduct) of 1.62 million tons U_3O_8 (in ore) at $15 per pound:[32]

> . . . Costs exceeding $10 a pound of U_3O_8 may be incurred by the early 1990s and . . . by the end of the century the cost of uranium could exceed $15 a pound. It is also possible that exploration world-wide will be sufficiently successful to avoid such costs.[33]

In summary, the 1971 outlook was characterized by naive technological optimism.

Concern about Cost Competitiveness of Nuclear Fuel

Two years later, the official estimates of cumulative uranium requirements through 2000 increased dramatically, to a range of 2.4 to 3.1 million tons U_3O_8. The sum of reserves plus byproduct plus estimated additional resources was 1.61 million tons U_3O_8 (in ore) at $15 per pound, that is, the same as before.[34] The nuclear capacity forecasts reached 825 to 1,500 GWe by the year 2000.[35] None of the increase in projected requirements can be explained by an increase in nuclear capacity. One must conclude that the outlook for gas reactors and breeder reactors darkened considerably. Warnings of a uranium shortage were sent to industry:

> Uranium requirements will aggregate substantially more than presently estimated United States ore reserves, potential resources and production

capability. Additional resources undoubtedly exist, but the lead times for discovery and construction of production facilities are long. A rapid expansion of the exploration effort is necessary in the near future if demands are to be met.[36]

In early 1973 the basic reason why the Atomic Energy Commission officials felt alarmed about the possibility of a uranium shortage was that they anticipated low prices for fossil fuels during the 1970s. The actions of OPEC, for example, were not foreseen. Consequently, the Commission believed that uranium prices would have to be kept at a very low level in order to maintain the cost competitiveness of nuclear power. However, very large quantities of uranium were known to exist in Chattanooga shale. These resources were considered appropriate for use in advanced breeder reactors, but not in light-water reactors. In April 1973 the cost of recovering uranium from 60- to 80-ppm shale was estimated to be $50 per pound of U_3O_8; a resource of 5 million tons U_3O_8 was estimated, for which the recovery rate was uncertain.[37] However, prior to the increase in OPEC oil prices, a price of $50 per pound of U_3O_8 was considered exorbitant—a threat to the competitive standing of nuclear power vis-à-vis fossil fuels.

In October 1972 the price of uranium for immediate delivery in the United States was $5.95 per pound U_3O_8.[38] At the same time, a report to the Edison Electric Institute forecast stable real prices: "Uranium prices will not increase faster than inflation, and if construction cost increases can be held down, nuclear power should gain an increased economic advantage over fossil fuel-based power."[39] Producers of uranium generally disagreed with this price assessment.[40] The Atomic Energy Commission also saw the need for a real price increase, but was not certain, as of March 1973, whether forward costs as high as $15 per pound U_3O_8 would have to be paid in order to clear the market:

> Uranium at costs of $15 or more per pound may well be economically competitive in water reactors. . . . The additional cost to the consumer of uranium from low-grade ore should be avoided if possible. In 1985 the difference between $8 and $15 uranium would amount to about $1 billion per year.[41]

In this context, reliance upon $50 uranium would have been regarded as a very rash suggestion.

Prospects for Chattanooga Shale

From 1973 to 1975 the prospects for commercial use of Chattanooga shale became less and less favorable. In December 1974 Battelle–Pacific Northwest estimated the costs to be roughly $80 per pound of U_3O_8.[42] In 1975 Hans Bethe estimated that the domestic nuclear industry might eventually require 2.8 million tons U_3O_8 in concentrate from Chattanooga shale, at a cost of $150 per pound of U_3O_8.[43] Also, in 1975 the Deputy Chief for Uranium and Thorium Resources at the USGS had grave doubts about both the economic and environmental feasibility of mining shales. "It seems to me that the Chattanooga will be one of the last alternatives to be considered."[44] Finally, Robert Nininger of the Energy Research and Development Administration expressed skepticism about the use of this resource.[45] A recent report to the Electric Power Research Institute corroborates this view.[46]

In table 6–2 we present data which illustrate the rise in cost estimates for low-grade resources over the period from 1969 to 1977. Although some allowance for inflation would be necessary to make the figures strictly comparable, it is clear that a tremendous quantity of uranium has vanished from the category of resources available at costs below $100 per pound U_3O_8.

If the utilities had been truly concerned about the long-term viability of nuclear power, the change in the outlook for Chattanooga shale would have been a matter of widespread controversy. However, utilities were much more interested in short-term financial matters, which were upset by the announcement by Westinghouse Electric Corporation on September 8, 1975 that it would not deliver 32,747 tons of U_3O_8 under contract because of unanticipated price increases. This action did not affect the quantity of uranium available from the domestic resource base at all, and the amount in question represented less than 1 percent of the amount present in Chattanooga shale. Nevertheless, the prospect of paying $30 or more per pound of U_3O_8, rather than $10, made utility executives take interest in the question of whether or not uranium should be considered a scarce resource.

Proponents of the breeder reactor, which had been suffering from delays and cost overruns, found the Westinghouse controversy to be a "shot in the arm" for the federal government's research program:

> Without the breeder, a shortage of nuclear fuel seems inescapable within twenty-five years. In fact, last month Westinghouse Electric—the major supplier of nuclear fuel—announced that from now on it will restrict deliveries to its customers, even though this means abrogating contractual agreements.[47]

As we have noted, the Westinghouse announcement has nothing to do with a long-term shortage of nuclear fuel, in either quantity or cost; it involves a dispute over who will pay for nuclear fuel, and the price to be paid.

Table 6-2
Past Estimates of Domestic Uranium Resources Recoverable at a Forward Cost between $30 and $100 per Pound U_3O_8
(*in millions of short tons U_3O_8*)

Source	Date of Estimate	$30–$50 Increment			$50–$100 Increment		
		Reasonably Assured	*Estimated Additional*	*Total*	*Reasonably Assured*	*Estimated Additional*	*Total*
WASH-1201	1 Jan. 1969	5	3	8	6	9	15
WASH-1098	Dec. 1970			1.4–8.6			9
WASH-1201	Jan. 1971	4	2	6	4	5	9
WASH-1243	April 1973			5[a]			8[b]
Battelle report	Dec. 1974			1.085[c]			9.7[d]
Nininger speech	28 Sept. 1975			Limited			Not known
GJO-100(77)	1 Jan. 1977	0.16[e]	0.28[f]	0.44			

Source: U.S. Atomic Energy Commission, *LMFBR Demonstration Plant Program*, WASH-1201 (Washington: GPO, 1972), p. 375; U.S. Atomic Energy Commission, *Potential Nuclear Power Growth Patterns*, WASH-1098 (Washington: USAEC, 1970), p. 2–14; Robert D. Nininger, "Uranium Reserves and Requirements," in *Nuclear Fuel Resources and Requirements*, WASH-1243 (Washington: USAEC, 1973), p. 16; Battelle–Pacific Northwest Laboratories, *Assessment of Uranium and Thorium Resources in the United States and the Effect of Policy Alternatives*, PB-238 658 (Springfield, Va.: NTIS, 1974), p. 5.19; Robert D. Nininger, "Current Uranium Supply Picture," (paper presented at the Conference on Nuclear Power and Applications in Latin America, Mexico City, September 28–October 1, 1975, p. 3; U.S. Energy Research and Development Administration, *Statistical Data on the Uranium Industry*, GJO-100(77) (Grand Junction, Colo: USERDA, 1977), pp. 21, 43.

[a]Shale, 60–80 ppm.
[b]Shale, 25–60 ppm.
[c]Does not include shale.
[d]Includes 2.6 million tons from Chattanooga shale and 7.1 million tons from marine phosphorites.
[e]Reserves
[f]Probable resources

Interest in Lifetime Requirements

Just as the Westinghouse controversy erupted, the critics of nuclear power came forward with a new argument against its continued growth: the claim that duty factors actually achieved would be much lower than design objectives, and therefore a uranium shortage would result. Raphael G. Kazman calculated lifetime requirements for reactors installed through 1981 (114 GWe),[48] and M.C. Day published a similar calculation for reactors installed through 1985 (250 GWe).[49] The Edison Electric Institute was less concerned about duty factors, but looked at lifetime requirements of reactors installed through 2000 (800 GWe), with recycle.[50] The Energy Research and Development Administration considered lifetime requirements for 320 GWe of capacity at 0.20 percent enrichment tails, with no recycle.[51] The Federal Energy Resources Council evaluated lifetime requirements for 300 GWe.[52]

In 1973 the Atomic Energy Commission's warning of a uranium shortage was based on the sum of reserves plus byproduct plus estimated additional resources, all at forward costs of $15 per pound of U_3O_8. In 1974 an estimate of $30 reserves was published for the first time, and in 1975 an estimate of $30 potential resources followed, with the definition of the term expanded to include "possible" and "speculative" resources which had formerly been ignored.[53] As a result, total estimated resources more than doubled, from 1.61 million tons U_3O_8 to 3.76 million tons.[54] If total nuclear capacity projections had stayed the same, this increase would not have offset lifetime requirements—for example, 7.5 million tons U_3O_8 (in concentrate) for 1,500 GWe of light-water-reactor capacity. However, as we have seen in chapter 4, nuclear capacity projections have been declining. Assuming a 90 percent recovery rate and 5 tons U_3O_8 (in concentrate) per megawatt, 3.76 million tons (in ore) will fuel 677 GWe—a figure which lines up with the year 2000 projections made in March 1976.[55] Thus by that date the outlook for long-term coverage began to depend on whether or not "possible" and "speculative" resources were included in the assessment.

On June 15, 1976, the Federal Energy Resources Council stated: "Together, the two categories of probable resources and reserves, including byproduct, amount to 1.84 million tons of uranium oxide. These are the highest reliability portion of total United States resources and serve as a prudent resource base for planning nuclear power plant construction programs."[56]

We believe this to be a sensible policy—the exclusion of "possible" and "speculative" resource from planning assessments. On this basis, it can fairly be said that prospects for an inadequate domestic resource base were practically eliminated in April 1977. On April 25, Energy Research

Table 6–3

Estimates of the Domestic Uranium Resource Base Commonly Used in Relation to Reactor Capacity Planning

(in millions of short tons U_3O_8)

	As of 1/1/75	As of 1/1/76	As of 1/1/77
$30 reserves (in current dollars), other than byproduct	0.60	0.64	0.68[b]
$30 probable resources (in current dollars)	1.14	1.06	1.09[c]
Byproduct, through the year 2000	0.14[a]	0.14	0.14
Subtotal	1.88	1.84	1.91
Cumulative ore production since 1947	0.2701	0.2824	0.2964[d]
Total	2.15	2.12	2.21
$30 probable resources as a percentage of the total	53%	50%	49%

Source: U.S. Energy Research and Development Administration, *Statistical Data of the Uranium Industry,* GJO-100(77) (Grand Junction, Colo.: USERDA, 1977), pp. 22, 24, 42, 43; GJO-100(76), p. 17.
[a]This value has been estimated on the basis of values reported for 1976 and 1977.
[b]The $30–$50 increment is an additional 0.16 million ST U_3O_8.
[c]The $30–$50 increment is an additional 0.28 million ST U_3O_8.
[d]Out of this total, 0.0938 million ST U_3O_8 in concentrate was delivered to domestic buyers.

and Development Administration officials forecast a "low" case of 380 GWe installed by 2000: "The 'LOW' case essentially captured policies enunciated by the President, both in terms of total energy and nuclear energy demands."[57] At the same time, estimates of $50 reserves and probable resources were prepared on the basis of drilling data and associated geological information.[58]

As table 6–3 shows, the estimate of $30 reserves and probable resources was 1.91 million tons U_3O_8 (in ore), to which 0.44 million tons can be added by raising the forward cost ceiling to $50. The delivery of 93.8 thousand tons (in concentrate) can be taken into account by adding 0.10 million tons (in ore) to the resource total. Thus the maximum allowable lifetime ore requirements for 380 GWe can be calculated simply:

$$\frac{2.45 \cdot 10^6 \text{ ST } U_3O_8}{3.80 \cdot 10^5 \text{ MWe}} = 6.45 \text{ ST } U_3O_8/\text{MWe} \qquad (6.3)$$

Given our assumptions in chapter 4—a duty factor of 25 million kWeh/ST U_3O_8, a capacity factor of 60 percent, and a recovery rate of 90 percent—this appears to allow reactor lifetimes of 28 years. If fuel burnup continues to be well below design-basis burnup, the construction of additional enrichment plant capacity can be undertaken in order to lower the tails assay and thereby lower fuel use. In short, the domestic resource situation is not tight enough to suggest that it would be imprudent to install 380 GWe by the year 2000.

It should be noted that 1.37 out of the 2.45 million tons under consideration have never been discovered. At the $30 cost level, probable resources are approximately equal to cumulative production plus reserves plus byproduct (see table 6–3). If cumulative production had been based on a $50 forward cost (in 1977 dollars), the same relationship would probably hold at the $50 cost level. In our judgment, a doubling of known resources can be taken as a "prudent resource base" for calculating long-term requirements. However, it should be clear that considerable uncertainty is associated with these figures.

As we have observed, there are several public-policy questions related to the domestic resource base. We have chosen to cover only the first of these in detail. The issue of import restrictions will be discussed briefly in chapter 7. Federal programs for resource assessment were discussed in chapter 2. Since no new probable resources need to be identified to cover requirements for 380 GWe, it appears that no increase in federal funding of uranium exploration is necessary—in fact, we do not see an urgent need to cover more than 300 GWe. Moreover, we see no need to deploy breeder reactors to sustain nuclear capacity growth to the year 2000. We shall not attempt to determine whether breeders will become competitive in cost with other forms of electricity generation, such as coal-fired plants, when our uranium resource base is depleted.

Notes

1. U.S. Atomic Energy Commission, *Nuclear Fuel Resource Evaluation: Concepts, Uses, Limitations* (Grand Junction, Colo.: U.S. Atomic Energy Commission, 1973), pp. 7, 16.

2. J.F. Hogerton, C.H. Barnes, L. Geller, and D.R. Hill, *Uranium Data*, EPRI EA-400 (Palo Alto, Calif.: Electric Power Research Institute, 1977), p. 3-5.

3. U.S. Energy Research and Development Administration, *Statistical Data of the Uranium Industry*, GJO-100(77) (Grand Junction, Colo.: USERDA, 1977), pp. 41–42.

4. Robert P. Blanc, "U.S. Uranium Potential Resources: Fact or Fiction?" in *Mineral Resources and the Environment, Supplementary Report: Reserves and Resources of Uranium in the United States*, prepared by the Committee on Mineral Resources and the Environment (Washington: National Academy of Sciences, 1975), p. 166.

5. U.S. Energy Research and Development Administration, *Statistical Data of the Uranium Industry*, GJO-100(77), pp. 23, 34–39.

6. R.J. McWhorter, D.R. Hill, E.A. Noble, J.F. Hogerton, and C.H.

Barnes, *Uranium Exploration Activities in the United States*, EPRI EA-401 (Palo Alto, Calif.: Electric Power Research Institute, 1977), p. C-4.

7. U.S. Atomic Energy Commission, *Nuclear Fuel Resource Evaluation*, pp. 17–18.

8. R.T. Zitting, "Estimation of Potential Uranium Resources," in *Mineral Resources and the Environment, Supplementary Report: Reserves and Resources of Uranium in the United States*, prepared by the Committee on Mineral Resources and the Environment (Washington: National Academy of Sciences, 1975), p. 144.

9. McWhorter et al., *Uranium Exploration Activities in the United States*, pp. B-78, B-80.

10. Ibid., pp. B-77, B-78.

11. U.S. Energy Research and Development Administration, *Statistical Data of the Uranium Industry*, GJO-100(77), pp. 41–43.

12. Robert D. Nininger, "Current Uranium Supply Picture," paper presented at the Conference of Nuclear Power and Applications in Latin America, Mexico City, September 28 to October 1, 1975.

13. Ibid., p. 3.

14. *Nucleonics Week*, Special Report, June 3, 1976; Owe Carlsson, "The Ranstad Project and Other Swedish Projects and Possibilities," in *Uranium Supply and Demand*, edited by M.J. Spriggs and K.D. Casteel (London: Mining Journal Books, 1977), pp. 208–209.

15. John P. Holdren, "Uranium Availability and the Breeder Decision," *Energy Systems and Policy* 1,3 (1975):226.

16. U.S. Energy Research and Development Administration, *Statistical Data of the Uranium Industry*, GJO-100(77), pp. 65, 68.

17. M.A. Lieberman, "United States Uranium Resources—An Analysis of Historical Data," *Science* 192, 4238 (30 April 1976):433.

18. DeVerle P. Harris, "The Estimation of Uranium Resources by Life-Cycle or Discovery-Rate Models—A Critique," draft, September 1976, p. 20.

19. Robert J. Kalter, Wallace E. Tyner, and Daniel W. Hughes, *Alternative Energy Leasing Strategies and Schedules for the Outer Continental Shelf*, A.E. Res. 75-33 (Ithaca, NY: Dept. of Agricultural Economics, Cornell University, 1975), pp. 18, 23–25.

20. U.S. Energy Research and Development Administration, *Statistical Data of the Uranium Industry*, GJO-100(77), p. 59.

21. McWhorter et al., *Uranium Exploration Activities in the United States*, p. B-88.

22. R.D. Nininger and S.H.U. Bowie, "Technological Status of Nuclear Fuel Resources," paper presented at the Conference on World Nuclear Energy, Washington, November 15, 1976, pp. 4-6.

23. U.S. Energy Research and Development Administration, *Statistical Data on the Uranium Industry*, GJO-100(77), p. 82.

24. "Polaris Resources Finds Uranium Oxide in Utah in Volcanic Formation," *Wall Street Journal*, August 15, 1977.

25. McWhorter et al., *Uranium Exploration Activities in the United States*, p. B-25.

26. Hogerton et al., *Uranium Data*, pp. 4-19 to 4-38; DeVerle P. Harris, *A Survey and Critique of Quantitative Methods for the Appraisal of Mineral Resources* (Grand Junction, Colo.: USERDA, 1976); Milton F. Searl and Jeremy Platt, "Views on Uranium and Thorium Resources," *Annals of Nuclear Energy* 2 (1975):751–762.

27. Rafford L. Faulkner, "Outlook for Uranium Production to Meet Future Nuclear Fuel Needs in the United States," paper presented at Fourth United Nations International Conference on the Peaceful Uses of Atomic Energy, Geneva, September 6–16, 1971, p. 6.

28. This assumes plutonium recycle and a 30-year lifetime. U.S. Atomic Energy Commission, *Nuclear Fuel Supply*, WASH-1242 (Washington: USAEC, 1973), p. 2.

29. U.S. Atomic Energy Commission, *LMFBR Demonstration Plant Program*, WASH-1201 (Washington: USAEC, 1972), p. 375.

30. Faulkner, "Outlook for Uranium Production," pp. 6–7.

31. R.W. Bown and R.H. Williamson, "Domestic Uranium Requirements," *Uranium Industry Seminar,* GJO-108(77) (Grand Junction, Colo.: USDOE, 1977).

32. U.S. Atomic Energy Commission, *LMFBR Demonstration Plant Program*, WASH-1201, p. 375; Faulkner, "Outlook for Uranium Production," Fig. 7.

33. Faulkner, "Outlook for Uranium Production," p. 7.

34. U.S. Atomic Energy Commission, *Nuclear Fuel Supply,* WASH-1242, pp. 2, 5.

35. Ibid., p. 2; and U.S. Atomic Energy Commission, *Nuclear Power Growth 1974–2000*, WASH-1139(74) (Washington: USAEC, 1974), p. 2.

36. U.S. Atomic Energy Commission, *Nuclear Fuel Supply*, WASH-1242, p. 11.

37. Robert D. Nininger, "Uranium Reserves and Requirements," in *Nuclear Fuel Resources and Requirements*, WASH-1243 (Washington: USAEC, 1973) p. 16.

38. Nuclear Exchange Corp., "Significant Events in the Uranium Market: 1969–1976," *Nuclear News* (December 1976):46.

39. National Economic Research Associates, Inc., *Fuels for the Electric Utility Industry, 1971–85* (New York, 1972), cited in "EEI Fuels Report Foresees U_3O_8 Prices as Unlikely to Increase," *Nuclear Industry* (October 1972):47.

40. Nuclear Exchange Corp., "Significant Events in the Uranium Market," p. 48.

41. Nininger, "Uranium Reserves and Requirements," pp. 21, 23.

42. Battelle–Pacific Northwest Laboratories, *Assessment of Uranium and Thorium Resources in the United States and the Effect of Policy Alternatives,* PB-238 658 (Springfield, Va.: NTIS, 1974), p. 5.30.

43. Hans A. Bethe, "Fuel for LWR, Breeders and Near-Breeders," *Annals of Nuclear Energy* 2 (1975):766.

44. Frank C. Armstrong, "Alternatives to Sandstone Deposits," in *Mineral Resources and the Environment, Supplementary Report: Reserves and Resources of Uranium in the United States,* prepared by the Committee on Mineral Resources and the Environment (Washington: National Academy of Sciences, 1975), p. 107.

45. See note 12.

46. McWhorter et al., *Uranium Exploration Activities in the United States,* p. B-84.

47. Ralph E. Lapp. "We May Find Ourselves Short of Uranium, Too," *Fortune,* October 1975, 151.

48. Raphael G. Kazmann, "Do We Have a Nuclear Option?", *Mining Engineering* (August 1975):36.

49. M.C. Day, "Nuclear Energy: A Second Round of Questions," *Bulletin of the Atomic Scientists* (December 1975):57.

50. John F. Hogerton, Clifton H. Barnes, William A. Franks, and Robert J. McWhorter, *Report on Uranium Supply: Task III of EEI Nuclear Fuels Supply Study Program* (New York: S.M. Stoller, 1975), p. 35.

51. John A. Patterson, "U.S. Uranium Supply Position," statement before Connecticut Public Utility Control Authority, Hartford, Conn., March 9, 1976, p. 5.

52. Federal Energy Resources Council, *Uranium Reserves, Resources, and Production* (Washington: The White House, 1976), p. 8.

53. U.S. Energy Research and Development Administration, *Statistical Data of the Uranium Industry,* GJO-100(77), pp. 26, 41–42.

54. See note 34; and Patterson, "U.S. Uranium Supply Position," Fig. 2.

55. Edward J. Hanrahan, "Demand for Uranium and Separative Work," paper presented at Atomic Industrial Forum Fuel Cycle Conference '76, Phoenix, Ariz., March 22, 1976, p. 12.

56. Federal Energy Resource Council, *Uranium Reserves, Resources, and Production,* p. 6.

57. Edward J. Hanrahan, Richard H. Williamson, and Robert W. Bown, "The Changing Nuclear Picture: Uranium and Separative Work Requirements," paper presented at Atomic Industrial Forum Fuel Cycle Conference '77, Kansas City, Mo., April 25, 1977, postscript.

58. U.S. Energy Research and Development Administration, *Statistical Data of the Uranium Industry,* GJO-100(77), pp. 21-22, 43.

7

The Domestic Uranium Market

This chapter consists of three parts: first, a discussion of uranium prices; second, a review of current legal disputes involving uranium; and third, a description of market structure. The focus of the whole discussion is our need to develop uranium price projections in order to simulate the behavior of a firm involved in leasing uranium lands. Unfortunately, the uncertainties surrounding uranium supply and demand are so great that we cannot assert much confidence in any particular price projection.

Uranium Concentrate Prices

Alternative Bases for Uranium Pricing

There are four basic ways in which domestic uranium concentrate prices may be established:

1. Given a competitive world market without import or export restrictions, prices will be determined by worldwide supply and demand, as shown in figure 7–1.
2. Given a competitive world market subject to a prohibition on exports from and imports into the United States, prices will be determined by domestic supply and demand, as shown in figure 7–1.
3. Given a foreign cartel and no import or export restrictions, prices will be set by the cartel, as shown in figure 7–2.
4. Prices may be regulated in order to ensure that the uranium producer earns no more than an allowable rate of return on invested capital. In other words, uranium production may be regulated in the same way that electric utilities are regulated.

In the first three instances, the distribution of economic rent associated with domestic uranium lands depends on whether or not an excess profits tax is imposed. If such a tax is present, rent is distributed between the landowners, from whom producers acquire mineral rights, and the government, which collects the tax. If an excess-profits tax is absent, rent is distributed between the landowners and the producers. Under rate-of-return regulation, the rent equals the payment made by producers to landowners.

In practice, import and export restrictions may be partial, not total,

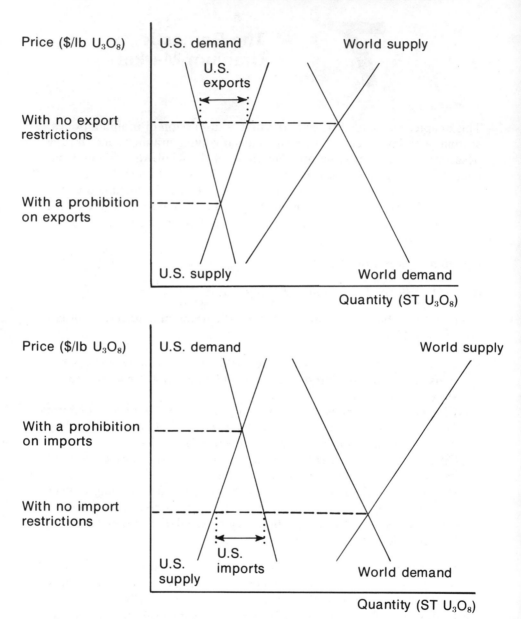

Figure 7–1. Competitive Market Behavior, in Two Possible Scenarios.

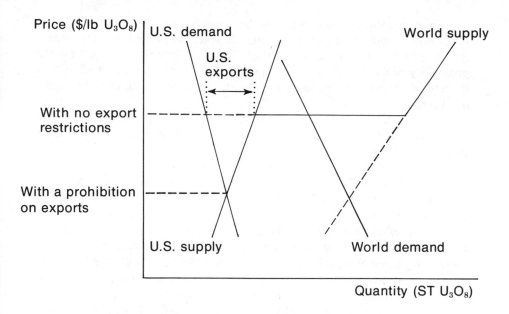

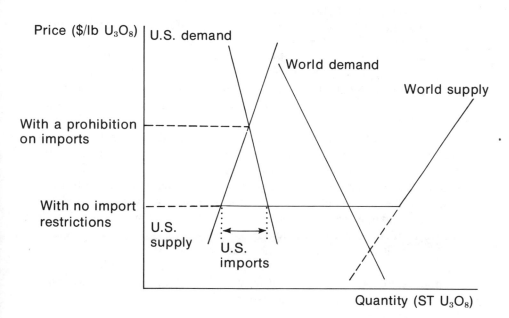

Figure 7–2. Influence of a Foreign Cartel, in Two Possible Scenarios.

and some uranium production may be subject to utility-style regulation while other production goes unregulated. Thus the preceding four categories may be blended in a single market. We have chosen not to address the possibility of a domestic cartel or a form of price regulation which differs from utility regulation because we consider these alternatives to be highly unlikely.

Within each of the three categories of unregulated prices, there are several different ways in which contracts can be written to establish U_3O_8 prices:

1. A contract for immediate delivery may specify a price to be paid in full at the time of delivery. In a competitive domestic market, this is a spot-market price, by definition. Uranium may be sold at auction,[1] through a brokerage firm, or at prices published by a brokerage firm.[2]

2. A contract for future delivery may specify a base price subject to escalation based on cost indexes to be published by the federal government or by some other third party. The base price may be divided into two or more portions, each of which is subject to a different escalation index; one such portion may not be subject to escalation at all.

3. A contract for future delivery may specify a fixed price, not subject to escalation or negotiation.

4. A contract for future delivery may specify the spot-market price prevailing at the time of delivery and published by a brokerage firm.

5. A contract for future delivery may stipulate that the price be negotiated in the future, and that delivery is not required if agreement on price cannot be reached. In effect, this sort of contract establishes a mutual option: the seller has an option to sell and the buyer has an option to buy.

6. A contract for future delivery may require that "baseball arbitration" is used to determine the price. Under this system, the buyer and seller send each other bids in sealed envelopes. If the difference in bids is no more than 5 percent, an average is taken; otherwise an arbitrator is called. On the next set of sealed bids, the arbitrator also puts a price in an envelope, and the price closest to his price is used whenever the difference in bids exceeds 5 percent.[3]

7. A contract for either immediate or future delivery may specify payment in some form other than full payment at the time of delivery. For example, some advance payment may be involved,[4] or stock ownership transferred, or mining claims transferred.[5]

Market-Price Contracts

A variant of the fifth type, negotiated price contracts, is the contract for future delivery at the market price in effect at the time of delivery, or at a particular time prior to delivery.[6] Sometimes a definite floor price is specified.[7] The difficulty with market-price terminology has been noted in connection with uranium royalties:

> . . . There is no "published" price which has received industry-wide acceptance. Nor is there a single figure which may legitimately be called a "market price."[8]

The practice of writing market-price contracts originated among an international cartel of uranium suppliers and spread to domestic suppliers.[9]

Utilities have publicly stated a preference for contracts for future delivery at base prices subject to escalation or at fixed prices.[10] If all domestic uranium suppliers refuse to offer uranium concentrates on these terms, one would have to suspect that some tacit or secret agreement was involved—an agreement which would constitute an antitrust violation. If domestic suppliers offered U_3O_8 at base prices subject to escalation, or at fixed prices, but quoted such high numbers that utilities elected to sign market-price contracts or some other alternative, the evidence of an antitrust violation would be relatively weak. Fixed-price contracts have been offered and signed at least as recently as October 1976,[11] but it is difficult to obtain information regarding current industry practice.

If a group of domestic uranium suppliers all offered market-price contracts subject to a precise definition—for example, "Nuexco Exchange Values" for immediate delivery, published by the Nuclear Exchange Corporation—one would have to suspect that some tacit or secret agreement was involved. Even though the prices in these contracts would not be fixed in the sense that specific numbers would be quoted in advance, the prices would be identical by definition. Moreover, if *market price* were uniformly interpreted as "spot-market price," one would suspect that the suppliers were planning to withhold sales and buy heavily in the spot market in order to drive up prices. As we have seen in chapter 4, the persistent tendency of actual industrywide deliveries to fall below scheduled deliveries suggests that the spot market represents a very small fraction of actual deliveries. The smaller this fraction, the easier it is for large firm or a group of firms to manipulate spot prices.

The "saving grace" of market-price contracts, as far as the antitrust laws are concerned, is their vagueness. One cannot prove that two or more market-price contracts will yield identical prices, given identical delivery dates. If it later turns out, however, that all these market-price contracts are interpreted in an identical fashion, evidence of an antitrust violation will be available.

The Department of Energy now offers a new form of uranium pricing which may be copied by one or more firms. On September 29, 1977 the Energy Research and Development Administration announced its terms of sale of enriched uranium:

> ERDA may furnish enriched uranium on a single transaction basis if the quantity of enriched uranium desired is small enough that undue effort would be involved in securing material through toll enrichment or if the customer demonstrates to ERDA's satisfaction that such action is appropriate because rapid delivery is required to meet an unforeseen emergency in which all reasonable attempts have been made without success, to procure natural uranium from commercial sources and, therefore, delivery of privately-owned natural uranium for toll enrichment is not possible. . . . The base charges for providing Government-owned uranium on a short notice, one time basis will be determined by utilizing a weighted average of prices paid during the previous year for natural uranium purchased under commercial contracts requiring payment of "market price at the time of delivery."[12]

Only the federal government is privileged to find out, through its annual survey of uranium marketing activity, what prices have been paid under every market-price contract. The definition of base charges is quite clear, and the particular value for a given year will, of course, be publicly available information.[13] If a group of domestic suppliers all offered to sell uranium at the government-published price, an antitrust violation would be suspected. However, one or two suppliers might offer this identical price without engendering much suspicion.

If the Department of Energy were to offer unlimited quantities of enriched uranium at the price terms just quoted, an effective ceiling on spot-market prices would be imposed, and spot-market sales might begin to represent a large fraction of total deliveries. Such actions would clearly be contrary to current policy, however.

Cost-Plus Contracts

A negotiated-price contract may require that the two parties agree on a price which ensures that the uranium producer earns no more than an allowable rate of return on invested capital. In other words, the contract may attempt to establish rate-of-return regulation without an independent regulatory authority. In December 1973 the Atomic Industrial Forum expected this sort of contract to be considered frequently as a method of protecting producers against risk:

> They have only two alternatives: quoting prices high enough so that they can be reasonably sure of making a fair profit under the most pessimistic production cost assumptions or negotiating what amount to cost-plus sales contracts that have much the same net effect.[14]

The difficulty with cost-plus contracts is that they may involve utilities in complicated disputes over the definition of "a fair profit," the capital investment on which this profit rate is to be earned and the operating costs which are to be covered by the contract. It may be much simpler for utilities to purchase or construct their own mining and milling facilities and earn an allowable rate of return set by a state regulatory commission. The role of the arbitrator in "baseball arbitration" may be similar to the role of such a commission.

Limitations on Escalation of Fixed-Price Quotations

Although it is difficult to make any definite statement about the way uranium prices will be determined in the future, a certain constraint on fixed-price quotations can be predicted on theoretical grounds. This constraint takes the form of a ceiling on the rate of escalation of fixed prices from one year of delivery to another, given a common date of quotation.

Elaborate security is required under Nuclear Regulatory Commission regulations for "special nuclear material"—uranium enriched in the isotopes U-235 or U-233 and plutonium. The principal purpose of these security requirements is to prevent the unauthorized construction of atomic bombs.[15] Enriched uranium can be derived from the natural uranium in yellowcake, with the use of an enrichment plant, and plutonium can be created with the use of a specially designed reactor and chemical processing plant. It is possible that in the future, these uses of yellowcake will be restricted by imposing security requirements on uranium concentrates and even uranium ores.[16] Such requirements could dramatically increase storage costs.

Under present regulations, uranium concentrates are stored under light security, and the cost of storage is little more than the financial cost of holding an expensive inventory. Yellowcake is typically stored in 55-gallon drums, which can be stacked without any danger of setting off a chain reaction, and which do not require radiation shielding. The value of U_3O_8 is roughly \$40 per pound, or \$80,000 per short ton—in other words, about 4,000 times the value of coal.

It follows that the value of advance deliveries of uranium can be measured simply by a present-value calculation. A buyer of uranium concentrates will consider purchasing uranium for delivery in every year, up to and including the year in which the uranium is needed to fulfill enrichment contract commitments. If the present value of the price quoted for a particular year is lower than the present value for other years, purchases will be made in that particular year. The discount rate for such calculations is the buyer's pretax cost of capital associated with nuclear fuel financing. We assume here that the buyer's pretax cost of capital is no greater than the producer's. Utilities generally have a lower cost of capital than mining companies, and so it is the utilities who stockpile uranium concentrates. If we assume that a group of buyers experience the same pretax cost of capital, then the rate of escalation of fixed prices quoted on a given date can be no greater than this discount rate. This situation is illustrated in figure 7–3.

If a shortage of uranium is known in advance—for example, if production appears certain to fall short of requirements in a given year in the future—prices will rise immediately for every year preceding the year of shortage, until the present value of the price in any preceding year is at least as high as the present value of the price in the year of shortage. This

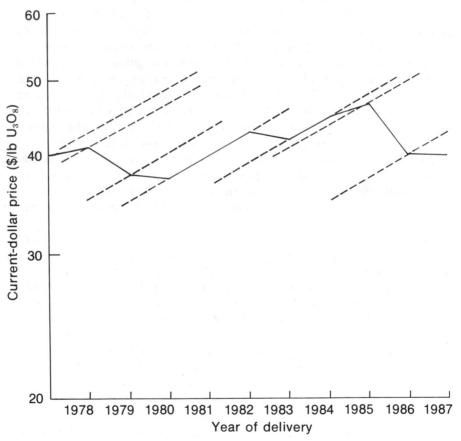

Note: Assuming a constant cost of capital, each dotted line represents a constant present value.

Figure 7–3. Example of a Set of Price Quotations Subject to a Limitation on the Rate of Escalation.

effect may lend a certain stability to uranium prices. Although the outlook for the next few years may change suddenly, prices quoted on a given date should not involve a sudden jump from one year of delivery to the next.

Limits to the Range of Uranium Concentrate Prices

Since many of the parameters involved in the nuclear industry are uncertain, it is common to bracket the range of possible values of a parameter by upper- and lower-limit calculations. We can describe U_3O_8 prices in these terms, but the upper and lower limits are so far apart that the usefulness of this sort of discussion is questionable.

In theory, a market may involve a variety of possible supply and demand curves, and yet prices may be constrained to lie between a price floor and a price ceiling (in the absence of government regulation). This situation is illustrated in figure 7–4. At the price floor, supply is perfectly elastic; at the price ceiling, demand is perfectly elastic. One possible

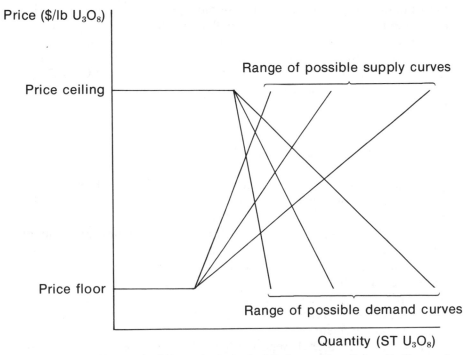

Figure 7–4. Theoretical Description of a Market with a Price Ceiling and a Price Floor.

cause of the kink in the demand curves would be the availability of a high-cost substitute for the commodity in question.

In the case of a worldwide market for uranium concentrates, the minimum price necessary to keep suppliers in business must be in the neighborhood of the November 1972 spot price, which was $5.95 per pound U_3O_8 (in 1972 dollars).[17] Production from very high-grade deposits could conceivably be sold at still lower prices. Instead of assuming a perfectly elastic portion of the supply curve, as in figure 7–4, we may consider the November 1972 market to be representative of the "worst possible" conditions suppliers are likely to face in the future. We assume that uranium demand may be severely curtailed by a halt to reactor construction, but it will not be wiped out entirely by a halt to worldwide reactor operation.

A price ceiling for U_3O_8 is probably determined by the domestic price of oil. If the cost per kilowatt-hour of U_3O_8 rose to the marginal cost per kilowatt-hour of oil purchased by U.S. utilities, it is likely that nuclear plants would be withdrawn from baseload service, and oil-fired capacity currently used for peaking would be returned to baseload service. In fact, this would probably occur if the cost per kilowatt-hour of nuclear fuel rose to that of oil. Rather than estimate minimum possible conversion, enrichment, and fuel fabrication costs, we will simply compare U_3O_8 costs with oil costs. In chapter 4 we estimated that a $1 per barrel of oil price is equivalent to $20.04 per pound of U_3O_8.[18] If we assume that oil prices fall to $10 per barrel, in 1977 dollars, then the price ceiling for U_3O_8 is $200.40 per pound, in 1977 dollars. At $20 per barrel, the ceiling would be $400.80 per pound.

Outlook for New Domestic Uranium

Since our interest is in prices faced by domestic producers leasing federal lands, we need not project prices for "old" uranium—that which will be delivered under fixed-price or base-price-plus-escalation contracts signed prior to 1976. However, price data for "new" uranium are relatively scarce. Recent uranium procurement has primarily consisted of market-price contracts and vertical integration by utilities. Figure 7–5 shows the percentage distribution of domestic uranium delivery commitments by type of procurement, as of January 1, 1977. It is clear that in future years, contract prices will represent a smaller and smaller percentage of total domestic procurement.

In October 1978 DOE reported that "there has been relatively little buying activity over the past $1\frac{1}{2}$ years, and prices for new procurement have remained stable over the past $2\frac{1}{2}$ years."[19] The short-term outlook

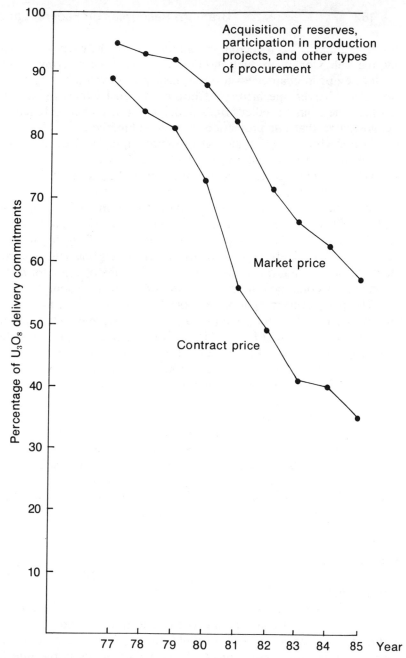

Note: These data describe firm delivery commitments for domestic buyers from domestic sources.

Figure 7–5. Types of Procurement, as of January 1, 1977.

for uranium prices is a decline in real terms. For the purposes of simulating the behavior of production complexes over lease terms—a task we shall attempt in chapter 9—it is necessary to develop a long-term price projection. Given the history of boom and bust cycles in the uranium market, one cannot predict prices with a great deal of confidence, and we acknowledge that our projection is quite subjective.

In our view, most studies of uranium supply and demand over the next 20 or 30 years tend to obscure a basic "catch 22" aspect of the market. If public opinion and government policy tend to favor nuclear energy as a substitute for fossil fuels, the demand for uranium will be very strong; if nuclear energy is out of favor, the supply of uranium concentrates is likely to be restricted. In either case, prices are kept at high levels. We do not think that the mining industry is fully cognizant of the challenges it is likely to face in the future regarding land use and the disposal of mill tailings. Delays and cancellations of uranium mines and mills may become common in the late 1980s, if not sooner.

The past trend in spot-market prices is shown in figure 7–6. These prices rose to $41.00 per pound U_3O_8, as of September 1976.[20] In October 1977 the Department of Energy's Chief of Uranium Supply estimated that market-price contracts for 1985 delivery would be fulfilled under prices of $31 to $44 per pound U_3O_8, in 1977 dollars ($50 to $70 per pound, in 1985 dollars).[21] With these figures in mind, we may speculate on the long-term trend in uranium prices.

Given the adequacy of the domestic uranium resource base, described in chapter 6, we expect the principal cause of upward pressure on real domestic prices to be cost increases associated with labor costs, fuel costs, and environmental and safety regulations. It is our subjective judgment that the real price of "new" domestic uranium (in 1977 dollars) will probably follow a long-term trend, from 1977 through 2000, of a 2 percent annual increase from a base of $40 per pound U_3O_8 in 1977.

Current Legal Disputes

The rise in spot-market prices illustrated in figure 7–6 was accompanied by a rise in prices quoted for future delivery. The primary focus of current legal disputes regarding uranium is the question of who should pay for U_3O_8 at "old" prices and who should pay for it at "new" prices, given various contractual obligations.[22] A secondary focus is the question of whether or not Gulf Oil, Rio Algom Mines, and any other U.S. corporations involved in the international uranium cartel should pay treble damages for price fixing.[23] Our interest is somewhat different: we would like to know whether recent prices have been artificially high as a result of actions

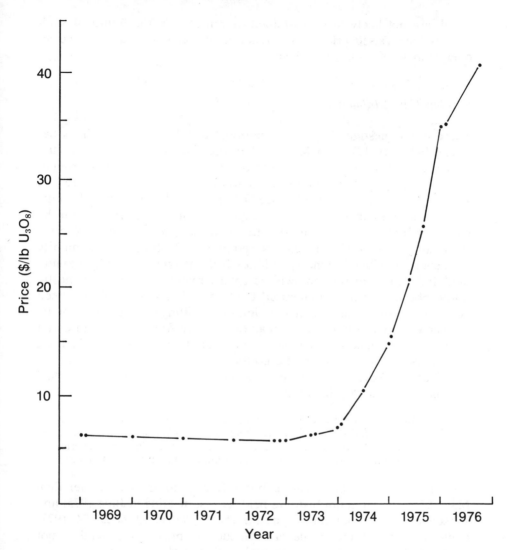

Source: Nuclear Exchange Corporation, "Significant Events in the Uranium Market: 1969–1976," *Nuclear News* (December 1976):46–51; Paul L. Joskow, "Commercial Impossibility, the Uranium Market and the Westinghouse Case," *Journal of Legal Studies* 6,1 (January 1977):137.

Figure 7–6. Nuexco Exchange Values for Immediate Delivery.

which are not likely to be sustained or repeated in the future. If such actions have raised prices, the preceding projection is likely to be an overestimate of future price levels.

Foreign Cartel Influence

There is no question that an international cartel met several times between February 1972 and May 1974 in order to fix prices for uranium concentrates.[24] In addition, meetings of Australian and Canadian producers were held. A basic chronology is shown in table 7–1. As figure 7–2 shows, a foreign cartel will influence U.S. prices unless import and export restrictions are imposed by the U.S. government. No restrictions on the export of U_3O_8 have been imposed, to our knowledge, although a Nuclear Regulatory Commission license is required.[25] Restrictions on imports effectively prohibited domestic utilities from utilizing imported uranium until 1977; these restrictions will be totally removed in 1984. Domestic utilities began purchasing imported U_3O_8 in 1970, and have always been permitted to stockpile it in anticipation of the lifting of restrictions on its enrichment for use in domestic reactors.[26] Therefore, the foreign cartel would have influenced domestic prices, even if cartel members had not quoted cartel prices to domestic utilities.

There is some evidence that the cartel broke up in 1974. It had difficulty preventing price cutting:

> . . . The battle to enforce the cartel's rules almost became physical at a January 1974 operating committee meeting in Johannesburg, according to an account by one participant, over a Rio Tinto gambit that several members considered to be "a very, very underhanded type" of deal.[27]

To our knowledge, there is no public reference to any cartel later than May 1974. The Uranium Institute held its first meeting in June 1975, and subsequently held conferences on June 15–17, 1976 and June 22–24, 1977. However, the Institute has never been sued for price fixing and does not appear to have been involved in such activities.[28]

If we accept the hypothesis that the cartel fell apart in 1974, then price increases since 1974 cannot be attributed to the cartel, and we are not in danger of overestimating future prices. If the cartel still exists, it is likely to continue to exist. The only way to show that we have overestimated future prices is to demonstrate that the cartel has survived up to the present, but will not survive much longer, and that it brought uranium prices up to the level of $40 per pound U_3O_8, in 1977 dollars. These two points are very difficult to substantiate. The forces which tend to act against a cartel have already been present; there is nothing really new

Table 7-1
Meetings of the Uranium Marketing Research Organization and Other Groups of Producers

Date		Location of UMRO Meetings	Location of Australian Meetings	Location of Other Meetings
	1971		Canberra[b]	
8 Dec.	1971			Canada[c]
1-4 Feb.	1972	Paris		
28 Feb.	1972		Canberra	
2 Mar.	1972		Sydney	
13-14 Mar.	1972	Paris		
20-21 Apr.	1972	Paris		
9 May	1972			Canada[d]
June	1972	Johannesburg		
6 July	1972	Cannes[a]		
7 Aug.	1972		Sydney	
2-3 Oct.	1972	Sydney		
1 Nov.	1972		Sydney	
17 Nov.	1972	Toronto		
9-11 Apr.	1973	Paris		
2-3 Aug.	1973	Johannesburg		
5,8-9 Oct.	1973	London		
21-23 Nov.	1973	Las Palmas		
28 Jan.- 1 Feb.	1974	Johannesburg		
May	1974	London		

Source: Except as noted, these meetings are listed in William F. Haddad, "Confidential Report to Irwin J. Landes, Chairman, Corporations, Authorities and Commissions Committee" (May 20, 1977, revised), published in U.S. Congress, House, Committee on Interstate and Foreign Commerce, Subcommittee on Oversight and Investigations, *International Uranium Cartel,* Hearings, 95th Cong., 1st Sess., May 2, June 10, 16, 17, August 15, 1977 (Washington: GPO, 1977), p. 665.

[a]Byron E. Calame, "Uranium Cartel Members Often Busy Protecting Interests From Each Other," *Wall Street Journal,* June 23, 1977, p. 14. According to William Haddad, the meeting was in Paris.

[b]The governments of Australia and Canada met in 1971; the Canadian minister for Energy, Mines, and Resources recommended "a world supplier-price for uranium around $6/lb." Zuhayr Mikdashi, *The International Politics of Natural Resources* (Ithaca, N.Y.: Cornell University Press, 1976), p. 108.

[c]The idea of having Gulf Minerals Canada join the cartel was suggested at a meeting between N. M. Ediger and Canadian government officials. Tim Metz and Byron E. Calame, "Gulf Oil Now Faces New Heat Over Role in Uranium Schemes," *Wall Street Journal,* April 25, 1977, p. 1.

[d]Canadian producers, including N.M. Ediger of Gulf, agreed in principle on the need for price fixing. Ibid.

about the incentives for price "cheating." Second, there is some evidence
to suggest that prices would not have been very different in the absence of
the cartel.[29]

In summary, we are doubtful that real uranium prices will fall in the
future as a result of a breakup of the cartel; even if the cartel still exists, it
probably is not responsible for the high level of current prices.

Possibility of a Domestic Cartel

There are at least seven pieces of evidence which tend to suggest that
domestic uranium producers have been involved in some sort of activity
which has restrained competition.

1. In 1972 utilities began to have difficulty finding suppliers who
would answer invitations to bid. At the Uranium Seminar sponsored by
the Atomic Industrial Forum on March 27, 1973, utility executives all
complained of the scarcity of bid responses.[30] Domestic producers and
buyers in 1972 were both aware of the fact that representatives of France,
Canada, Australia, and South Africa had met "to end their price warfare"
and create an "arrangement that would put a floor under overseas
prices."[31] This news provided an incentive for suppliers to wait for price
increases. While it would not be surprising to find some bids much higher
than others, it is odd that so many suppliers offered no bids at all.

2. In February 1973 the Atomic Industrial Forum convened a meeting
in San Diego which was sufficiently "sensitive" to discourage Exxon
from taking part. A Gulf Oil executive described the situation in an
internal memorandum dated February 23:

> Tom Upchurch advised me that he had come down to Los Angeles on
> Monday to attend the AIF meeting to discuss the embargo question
> where other producers were present. Exxon has very conservative
> guidelines on such participation and when Tom discovered that the AIF
> representation was not acceptable (lack of a lawyer and/or M. Gordon),
> he elected not to participate and returned to Seattle.[32]

3. Uranium producers met at Oak Brook, Illinois in March 1973 at a
meeting to which government officials were not invited. A representative
of the Department of Energy has described it as follows:

> The Oak Brook meeting and most other meetings of the Atomic Indus-
> trial Forum usually had a meeting of the Mining and Milling Committee.
> This would be a private meeting just for the members of the committee. It
> would be essentially uranium producers, domestic and foreign. This
> would not have been a public part of the proceedings, and I would not
> have attended.[33]

4. Following the Oak Brook meeting, one of the participants in the international cartel—a representative of Rio Tinto Zinc—wrote to the president of a U.S. producer about the advantage of market price contracting:

> As we have not had the opportunity before of discussing this concept with a U.S. producer I wondered whether the enclosed example of how we operate the world market concept might be of interest to you in order that perhaps we could discuss this together when we next meet.[34]

5. In the fall of 1973, suppliers continued to be reluctant to answer utility invitations to bid.[35] Many utilities found this frustrating.[36]

6. By March 1974, domestic producers were issuing identical price quotations. Erik Kvaven, a representative of the Tennessee Valley Authority, later described the situation as follows in congressional testimony:

> *Mr. Kvaven.* When George White of Nuexco started publishing prices, every time he published a new price, that was the new quote that you got from everybody.
>
> *Mr. Maguire.* Mr. Chairman, excuse me. If that is true, Mr. Kvaven, does that not lead to a suspicion?
>
> *Mr. Kvaven.* It leads to a suspicion that they all read Nuexco.
>
> *Mr. Gore.* And they agreed not to undercut each other.
>
> *Mr. Kvaven.* Evidently. . . .
>
> *Mr. Gore.* . . . It seems to me that if everybody is quoting the same price and nobody is offering to sell at a somewhat lower price in order to get customers, then there is no competition and something is wrong with the marketplace.
>
> *Mr. Kvaven.* That is right. It is a complete seller's market right now.
>
> *Mr. Gore.* But it was a buyer's market before the producers refused to sell?
>
> *Mr. Kvaven.* Yes.[37]

7. In the fall of 1974, utilities reluctantly turned to market-price contracts. "The latter part of 1974 also witnessed another milestone in the domestic market, namely the first large-scale buyer acceptance of unpriced future deliveries."[38] As we have noted earlier, uniform interpretation of market-price contracts would involve identical prices. These contracts have become increasingly common. Following its July 1, 1975 price survey, the Energy Research and Development Administration began surveying the proportion of supply commitments covered by market-price contracts.[39]

Taken together, these bits of information suggest that the competi-tiveness of the domestic uranium industry may have been restrained. However, more definite evidence would be necessary to prove an anti-trust violation.

Westinghouse "Short" Position

"Short" sales of uranium consist of contracts for future delivery signed by companies which cannot produce the amount in question from their own reserves in time to make the scheduled deliveries and have not contracted with producers to make up this deficit. Thus "short" sales create excess demand for future deliveries. The most well-known "short" is that of Westinghouse Electric Corporation:

> On July 14, 1975, Westinghouse announced that it was short between 40 and 60 million pounds of uranium for the period 1978–1995. Outside estimates have put that short position at close to 70 million pounds of uranium concentrate.[40]

On September 8, 1975, Westinghouse notified utilities in the following terms that it did not intend to fulfill its uranium supply contracts:

> The purchase of uranium in the open market at current prices would involve such an unfavorable and large burden as to be "commercially impracticable" from a legal viewpoint.[41]

This announcement was followed not only by litigation, but by increases in procurement activity by utilities and a continuing rise in the spot-market price of uranium.

The sudden realization by utilities that they might have to purchase 40 to 60 million pounds of U_3O_8 from producers or produce it themselves must have caused prices to rise. However, if demand were inelastic, the resulting prices could not be considered artificially high. The market moved from a position of excess demand, with artificially low prices, to a position of equilibrium, with normal prices. We know, of course, that demand was actually elastic; otherwise Westinghouse would not have found it necessary to offer U_3O_8 at low prices in order to persuade utilities to buy reactors. If nuclear plants on order had not been delayed or cancelled, we could conclude that current prices are artificially high. Utilities were misled by Westinghouse's low uranium price quotations to order more nuclear capacity than they would have in the absence of such offers. We know, of course, that delays and cancellations have been widespread, and so it is doubtful that such artificially high demand still exists. However, even if current nuclear capacity growth has been inflated by Westinghouse's uranium sale contracts, further delays and cancellations in connection with uranium prices seem unlikely. Prices may be artificially high, in which case they will continue to be artificially high, and our projection is still valid.

The objection may be raised that recent prices simply reflect a temporary reaction to the Westinghouse affair: a large amount of uranium had to be ordered all of a sudden, and the timing of these orders, not the quantity, has created an artificial price increase. This argument is not very persuasive because it has been over 3 years since the announcement was made. A temporary jolt should have been worked out by now. Finally, the claim is sometimes made that the Westinghouse "short" made utilities panic over uranium availability and try to obtain much longer-term coverage than is really necessary. This argument, too, is unconvincing, given the data in chapter 4. Coverage for more than nine reloads is rare. Present levels of coverage would probably have been sought if utilities had been forced to procure uranium without help from Westinghouse.

There is a very strange aspect of the uranium market which ought to be noted here. While nuclear-capacity projections were falling, prices for future delivery were rising. Normally, a drop in demand should cause a drop in prices; under such conditions, Westinghouse's "short" sales would have had the effect of delaying purchases to a time when prices had fallen. This did not happen because of the peculiarities of the uranium market.

In summary, consideration of the Westinghouse controversy does not suggest that we have mistaken a temporary price fluctuation for a permanent increase. The willingness of Westinghouse to sell uranium at low prices, despite the uncertainties surrounding the uranium market, demonstrates that Westinghouse used uranium as a loss leader. Even though the firm's executives might have believed that the most likely trend in uranium prices would be in line with the long-term contracts they signed, it must have been clear that Westinghouse was accepting a substantial risk. Turnkey contracts for nuclear plants were also used as loss leaders to get the commercial nuclear industry through a "takeoff" phase.

In this context, one might ask whether the mining and milling companies sold uranium as a loss leader. It is important to note that during the early 1970s the mining industry was perennially complaining that it lacked an incentive to explore for the quantities of uranium required to meet nuclear capacity projections. Therefore, if the industry believed these projections would be attained, or even approached, the industry could not have sold uranium so cheaply without knowing that future sales at higher prices were being foregone. In other words, the mining industry must have been bluffing about its hard times, while seeing a strong sellers' market in the years ahead.

We doubt that this is really what happened, because it appears that mining firms never really believed the Atomic Energy Commission's forecasts. Having been cheated by the AEC before, in the uranium bust of the early 1960s, these firms decided that stockpiling uranium for the 1980s market would be a foolish gamble. Utility companies apparently mistook this skepticism for evidence that uranium was plentiful and cheap.

Gulf "Long" Position

From March 1972 through March 1974, Gulf Oil Corporation and General Atomic Corporation (a fifty-fifty joint venture between Gulf and the Royal Dutch/Shell group of companies) purchased uranium concentrates for future delivery: 6 to 10 million pounds U_3O_8 from Ranchers Exploration and Development, over 11 million pounds from Exxon and Union Carbide, 5 million pounds from the Sohio/Reserve Oil and Mineral joint venture, and 27 million pounds from United Nuclear.[42] Since 1974 the General Atomic Corporation has cancelled plans to construct high-temperature gas reactors (HTGRs) and has not obtained enough light-water reactor (LWR) fuel contracts to make use of all this purchased U_3O_8. In 1973 Gulf purchased properties at Mt. Taylor, New Mexico, which increased its reserves in that area, but so far none of the Mt. Taylor reserve has been committed or sold, despite its vast size.[43] "Gulf alone owns the largest single body of uranium ore reserves in the United States, at Mt. Taylor, New Mexico, estimated by Gulf in 1976 to be at some 135 million pounds."[44] Thus General Atomic and Gulf are both in a "long" position; their reserves and purchase commitments exceed their delivery commitments. The Nuclear Exchange Corporation estimated that as of January 27, 1977, General Atomic's delivery commitments were low enough to leave an inventory of 17 million pounds in 1985, not counting production from Gulf's reserves.[45]

Although Gulf may have originally acquired reserves in order to sell HTGRs with U_3O_8 included in the fuel contracts, it is clear that Gulf is now withholding uranium from the present LWR market in order to sell it in future LWR markets. If Gulf held a small share of domestic supplies, it would simply be a speculator, unable to influence uranium prices. However, the magnitude of Gulf's involvement requires that we consider the matter differently. If we assume that the domestic uranium market is competitive, then it appears that prices have risen, in part, as a result of the decision by Gulf and General Atomic to withhold U_3O_8 from utilities, and that prices will be lower when this material is eventually sold than they would otherwise have been. In other words, these companies have created a present scarcity and a future surplus by shifting a portion of the supply available to utilities from the present into the future. Our uranium price projection, based largely on past spot-market trends, may be an overestimate.

The possibility of a reduction in price increases, or even a reduction in prices, becomes clear when we examine General Atomic's purchase prices. The Ranchers Exploration and Development contract specified $7.85 per pound U_3O_8 for 1976 delivery, and $9.05 for 1980 delivery; the United Nuclear contract specified $8.46 for 1978 delivery. At these prices, General Atomic could afford to lower U_3O_8 prices by "dumping" concentrates on the market; the sale of this material would still be highly profitable when compared with its cost to General Atomic. Similarly, Gulf would be able to "dump" its Mt. Taylor reserves and still make a handsome profit, if they were acquired sufficiently cheaply.

According to United Nuclear's brief against General Atomic, the Nuclear Exchange Corporation has estimated that as of January 27, 1977, General Atomic's cumulative purchase commitments through 1985 exceed its cumulative sales commitments through 1985 by 17 million pounds, or 8,500 short tons. This amounts to only about 3.7 percent of the sum of 1977 deliveries plus apparent 1978–1985 buyer requirements.[46] However, the amount of reserves that General Atomic decided not to commit for delivery, but could have committed, is likely to be more significant. The total amount withheld from long-term contracts may be 10 percent of the total market. Of course, we do not know whether all the deliveries to General Atomic will actually be made, since they are subject to litigation. Moreover, we do not know how other firms would have behaved if they had owned the uranium under General Atomic's control. We believe that the net effect of the "long" position will not be great enough to require an alteration of our projection of long-term price trends.

Market Structure

Independent Mines

The preceding discussion has been based on prices for U_3O_8 in yellowcake. It is possible, however, for uranium produced from leased federal lands to be sold in the form of ore. Although our cost data pertain to integrated mine-mill complexes (see chapter 5), it is important to recognize the existence of independent mines, that is, those which are not owned by the company which mills their ore.

Historically there has been a trend toward vertical integration of uranium mining and milling, but this trend may be reversed in the future. In 1955, 35.0 percent of milled ore was "captive ore," that is, mined by the milling company. This percentage rose to 64.0 percent in 1960, 73.0 percent in 1965, and 95.5 percent in 1970.[47] The primary explanation for this trend lies in the AEC's change in procurement policy. During the early 1950s, the AEC purchased ore according to a price schedule which subsidized high-grade ore production at the expense of low-grade ore production. Small, remote mines with high-grade ore were made profitable. Haulage allowances were provided by the AEC in order to subsidize mines in remote areas.[48] However, after March 31, 1962, the AEC purchased no ores and limited its yellowcake purchases primarily to production from reserves which existed on November 24, 1958.[49] Small mining companies, which had often chosen to risk exhausting their ores rather than spend the capital required to prove reserves, protested the AEC's policy. On February 21, 1961 the AEC made a modest concession to these companies and agreed to award certain contracts according to historical production levels.[50] AEC uranium procurement ended in 1970.

Recent evidence suggests that independent mining companies may be making a comeback. The famous Hidden Splendor Mine, which was shut down in the late 1950s, has reopened.[51] Some utilities have purchased rights to uranium production from independents.[52] Most of the companies conducting exploration drilling in 1975 had not announced plans to construct mills.[53] Perhaps the strongest evidence for the viability of independent mining ventures is the fact that Gulf Oil is in the process of investing about $400 million in what is "expected to be the largest and deepest uranium mine in the U.S." without building a mill.[54] No doubt Gulf's ability to build its own mill prevents other companies from charging exorbitant prices to process Gulf's ore. This is not much consolation to smaller companies, but it does show that vertical integration is not a necessary aspect of new uranium ventures.

One of the practices established by the Atomic Energy Commission to promote uranium supply during the 1950s was the creation of ore-buying stations, where independent miners could sell ore in small quantities on a spot market. This practice has been revived by General Electric and the Atlas Corporation, a uranium supplier.[55]

Vertical Integration by Utilities

As we have seen in figure 7–5, vertical integration by utilities is projected to absorb an increasing proportion of the domestic market in the future. If a utility owns its own mine and/or mill, the U_3O_8 price is likely to be subject to rate-of-return regulation. The regulatory problems associated with captive uranium mines are similar to those associated with captive coal mines.

September 1975 was a turning point in utility attitudes toward uranium supply activities. Westinghouse's announcement of "commercial impracticability" led to widespread acquisition of reserves by utilities. Prior to the announcement, few companies were involved, as table 7–2 shows. However, by March 1976 the Energy Research and Development Administration reported a new trend:

> As a consequence of the difficulties in contracting for uranium and rapid price movement, the interest of utilities in participating in uranium exploration and production projects has been increasing sharply. At least 13 domestic utilities are now involved in such projects.[56]

By April 1977 the number of utilities involved had doubled.[57] One of the risks faced by utilities, under the Mining Law of 1872, is that claims purchased may turn out to be fraudulent.[58] Controversy over this issue may lead to increased interest in replacement of the Mining Law of 1872 by a leasing system.

Vertical Integration by Conversion and Fuel Fabrication Firms

Another important consideration in the study of uranium concentrate prices is the existence of vertical integration, beyond uranium milling, by conversion and fuel-rod fabrication firms. In situations where a firm other than an electric utility produces yellowcake for its own use, there is no explicit price for uranium concentrate, and there is no basis for rate-of-return regulation. At present there are only three firms in this category: Kerr-McGee, which sells uranium hexafluoride; Exxon, which sells nuclear fuel rods; and Westinghouse, which sells both nuclear fuel rods and nuclear steam supply systems.

Conversion and fuel-rod fabrication are the only two portions of the nuclear fuel cycle, beyond uranium milling, in which forward vertical integration is likely to exist during the next decade. Privately owned enrichment plants are effectively precluded by the expansion of federal capacity,[59] and fuel reprocessing plants are unlikely to enter operation because of the unfavorable political outlook for plutonium use and the President's decision to offer "neither federal encouragement nor funding."[60] Fuel-rod fabricators are likely to continue to produce their own enriched uranium dioxide and pellets rather than buy from independent

Table 7–2

Direct Participation by U.S. Utilities in Uranium Supply, through July 1975

Utility	Participation in AEC/ERDA Survey of Exploration	Acquisition of Land and Mineral Rights	Participation in an Active Exploration Program	Acquisition or Construction of a Uranium Mill
Houston Natural Gas[a]		1968[c]	1968[c]	
Southern California Edison			1971[d]	1975[e]
Tennessee Valley Authority	1972[f]	1972[g]	1973[h]	1974[i]
Commonwealth Edison	1974[j]	1974[k]		1974[k]
Northern States Power	1974[j]		1971[k]	
Central Power and Light[b]	1975[l]			
Public Service Co. of Oklahoma	1975[l]			

[a]This is a natural gas utility, not an electric utility.

[b]Subsidiary of Central and South West Corporation.

[c]This joint venture was initially arranged with Combustion Engineering and Ranchers Exploration and Development Corporation, but Houston Natural Gas bought Combustion's 25 percent interest in 1972. See *Moody's Public Utility Manual, 1972.*

[d]"Southern California Edison Is First Utility in Uranium Exploration," *Nuclear Industry,* March 1971, p. 46.

[e]An environmental report on the proposed mill was filed with USNRC in July 1975. All mining claims, plus the preference right lease, are held by the Union Pacific Railroad Co. See U.S. Nuclear Regulatory Commission, *Final Environmental Statement Related to Operation of Bear Creek Project, Rocky Mountain Energy Company,* NUREG-0129 (Washington: USNRC, June 1977), p. 1-4.

[f]U.S. Atomic Energy Commission, *Uranium Exploration Expenditures in 1971 and Plans for 1972–73,* GJO-103(72) (Grand Junction, Colo.: USAEC, 1972), p. 7.

[g]"TVA Buys Stake in Uranium Claims," *Nuclear Industry,* April 1972, p. 39.

[h]Walter C. Woodmansee, "Uranium," in *Minerals Yearbook, 1973,* issued by U.S. Department of the Interior, Bureau of Mines (Washington: GPO, 1975), p. 1264.

[i]Walter C. Woodmansee, "Uranium," in *Minerals Yearbook, 1974,* (Washington: GPO, 1976), p. 1,332.

[j]U.S. Atomic Energy Commission, *Uranium Exploration Expenditures in 1973 and Plans for 1974–75,* GJO-103(74) (Grand Junction, Colo.: USAEC, 1974), p. 9.

[k]"Commonwealth Edison's Purchase of Cotter Sets Precedent," *Nuclear Industry,* May 1974, p. 32.

[l]U.S. Energy Research and Development Administration, *Uranium Exploration Expenditures in 1974 and Plans for 1975–76,* GJO-103(75) (Grand Junction, Colo.: USERDA, 1975), p. 11.

suppliers. Kerr-McGee discontinued its fuel pellet sales in 1975 when "a lack of additional contracts on terms that would permit continued profitable operations resulted in the facility being placed on a standby status, effective at the end of the year."[61]

There are two principal obstacles to vertical integration into conversion or fuel-rod fabrication: economic barriers to entry and federal merger policy. Conversion is characterized by major economies of scale, and there are only two firms in the domestic industry—Kerr-McGee and Allied Chemical. Fuel-rod fabrication requires a lot of technical expertise in reactor design, and only Exxon is presently able to compete with the four reactor manufacturers (see table 7–3). Federal merger policy is reflected by two recent mergers: the acquisition of Utah International by General Electric and the acquisition of Anaconda by Atlantic Richfield. In the first instance, the Justice Department required General Electric to set up an independent corporation. "The Uranium Subsidiary shall be formed for the sole purpose of engaging in the uranium business provided, that the Uranium Subsidiary shall be prohibited from selling or otherwise transferring any uranium to GE. . . ."[62]

In the second instance, the Federal Trade Commission filed suit on the grounds that the merger might "substantially lessen competition and unreasonably restrain trade in the production and sale of uranium oxide and copper."[63] Within 3 weeks, Atlantic Richfield announced the sale of its entire uranium holding to U.S. Steel Corporation.[64] Even then, the FTC opposed the merger on the grounds that Atlantic Richfield was a "most likely potential entrant" into the uranium concentrate market. The courts did not find this argument to be persuasive.[65]

Horizontal Divestiture and Similar Restraints

At present there exists substantial political support for the idea that major oil corporations should not be allowed to expand their uranium production facilities or uranium reserve ownership. The implication behind this concept is that oil companies, or diversified energy companies, are more likely to engage in anticompetitive activity in the uranium market than other sorts of companies. We do not find evidence for this hypothesis. Gulf Oil has participated in a uranium cartel, while Exxon refused to even attend a meeting in San Diego on the subject of imports. Rio Tinto Zinc is a cartel member, but Anaconda and Phelps Dodge are not. Thus we have discussed the possibility of activities in restraint of trade without speculating on some special role played by oil companies. It is necessary to recognize, however, that legislation which is likely to be ineffective as a restraint on price fixing may nevertheless have a major impact on prices.

Table 7-3
Extent of Vertical Integration of the Uranium Milling and Nuclear Fuel Fabrication Industries, 1977

Uranium Milling Companies with No Involvement in Nuclear Fuel Fabrication	Uranium Milling Companies with Limited Involvement in Nuclear Fuel Fabrication	Uranium Milling Companies Involved in Nuclear Fuel Fabrication	Reactor Manufacturers Involved in Nuclear Fuel Fabrication
American Nuclear	Atlantic Richfield[c]	Exxon	Babcock and Wilcox
Anaconda	Gulf Oil[d]	Westinghouse Electric[a]	Combustion Engineering
Atlas	Kerr-McGee[c]		General Electric[c]
Commonwealth Edison	Lucky Mc[e]		
Continental Oil	Union Carbide[f]		
Federal Resources	United Nuclear[c]		
Homestake Mining			
Intercontinental Energy[a]			
Mobil Oil[a]			
Niagara Mohawk Power[a]			
Newmont Mining			
Phelps Dodge			

Pioneer Natural Gas
Reserve Oil and Minerals
Rio Tinto Zinc
Standard Oil Co. of California[b]
Standard Oil Co. of Ohio
Union Oil Co. of California[b]
Union Pacific Railroad Co.
United States Steel[a]

[a]These companies operate in situ leaching facilities. Westinghouse Electric is a reactor manufacturer.

[b]These companies have announced plans to construct uranium mills.

[c]These companies have operated fuel fabrication facilities in the past. Atlantic Richfield and United Nuclear sold complete rods, while Kerr-McGee sold only fuel pellets.

[d]Gulf Oil has operated light-water reactor fuel fabrication facilities in the past, and presently operates high-temperature gas reactor fuel fabrication facilities. Gulf is constructing a uranium mine which is expected to be the largest and deepest in the United States. Gulf also operates a uranium mill in Canada, through a joint venture with a West German company. Uranerzbergbau. However, Gulf has not announced plans to construct a uranium mill in the United States.

[e]Lucky Mc is owned by General Electric. In 1976 GE acquired Utah International, which owned Lucky Mc. In giving its approval to the merger, the U.S. Department of Justice required that Lucky Mc be managed in such a way that it cannot be controlled by GE and cannot sell uranium to GE.

[f]Union Carbide operates the Oak Ridge National Laboratory for DOE. This laboratory is involved in research and development of nuclear fuels. However, Union Carbide is not involved in commercial nuclear fuel fabrication.

Any measure which limits the eligibility of companies to invest in uranium is likely to create an upward pressure on prices. This result would not occur, of course, if in the absence of such legislation all new uranium projects were owned by utilities.

On September 8, 1977, proponents of restraints on horizontal diversification by energy companies suffered a major setback. "By the surprisingly lopsided vote of 62 to 30, the Senate killed a proposal that would have prohibited any of the 16 largest U.S. oil companies from buying additional coal or uranium properties."[66] Accordingly, we do not anticipate that such measures will be passed in the near future. Moreover, we are not prepared to estimate the impact such legislation would have on uranium prices, except to note that it would probably shift a share of the market from competitive pricing to rate-of-return regulation.

Notes

1. "Utah International, Atlas Sell 5-Million Pounds U_3O_8 in March Auctions," *Nucleonics Week*, April 15, 1976.

2. The Nuclear Exchange Corporation publishes "Nuexco Exchange Values" for immediate delivery; the Nuclear Assurance Corporation publishes "Fuel-Trac" prices.

3. " 'Baseball' Arbitration Being Used More and More on Uranium Prices," *Nuclear Fuel*, November 1, 1976.

4. "Intercontinental Energy Schedules Sale of Uranium," *Wall Street Journal*, December 7, 1976, p. 17; "An Option on the Zamzow Lease," *Nuclear Fuel*, November 1, 1976; *Nucleonics Week*, June 24, 1976, p. 2.

5. "Western Nuclear Deal," *New York Times*, October 5, 1977.

6. John A. Patterson, "Status of Uranium Procurement," paper presented at the Atomic Industrial Forum Fuel Cycle Conference '77, Kansas City, Mo., April 25, 1977, p. 5.

7. John F. Hogerton, Clifton H. Barnes, William A. Franks, and Robert J. McWhorter, *Report on Uranium Supply*, Task III of EEI Nuclear Fuels Supply Study Program (New York: S.M. Stoller, 1975), p. 52.

8. Royal E. Peterson, "The Uranium Royalty Provision," *Rocky Mountain Mineral Law Institute* 22 (1976):873–878.

9. Hogerton et al., *Report on Uranium Supply*, p. 53; testimony of George White, President of Nuclear Exchange Corp., in U.S. Congress, Joint Committee on Atomic Energy, *Proposed Modification of Restrictions on Enrichment of Foreign Uranium for Domestic Use, Hearings,*

93rd Cong., 2d Sess., September 17 and 18, 1974 (Washington: GPO, 1975), p. 47; William Grieder, "Uranium Cartel: Litigation Tangle Follows Price Rise," *Washington Post*, May 8, 1977, p. A1.

10. For example, see U.S. Congress, Joint Committee on Atomic Energy, *Proposed Modification of Restrictions*, pp. 47, 100.

11. "Pennsylvania P & L Buys UNC Uranium at $56 in 1980 and $62 in '82," *Nucleonics Week*, October 28, 1976.

12. 42 F.R. 51636 (September 29, 1977).

13. Prices paid by government-owned utilities, such as the Tennessee Valley Authority, are also publicly available. Private contracts might conceivably involve prices which are tied to some government purchase price.

14. "Uranium Market Surge Opens New Era for Buyers, Sellers," *Nuclear Industry*, December 1973, p. 21.

15. 10 C.F.R., Part 70.

16. Paul Hofmann, "Escort Unit Urged for Uranium Cargo," *New York Times*, April 30, 1977; "Common Market Says It Increased Security on Nuclear Materials," *Wall Street Journal*, May 3, 1977, p. 12; Walter Sullivan, "Nuclear Safeguards Assessed at Meeting," *New York Times*, May 22, 1977; "Uranium: The Israeli Connection," *Time*, May 30, 1977, pp. 32–34.

17. This was the "Nuexco Exchange Value" for immediate delivery. Nuclear Exchange Corporation, "Significant Events in the Uranium Market: 1969–1976," *Nuclear News*, December 1976, p. 47.

18. See equation 4.9.

19. George F. Combs, Jr. and John A. Patterson, "Uranium Market Activity," *Uranium Industry Seminar*, GJO-108(78) (Grand Junction, Colo.: USDOE, 1978).

20. Nuclear Exchange Corporation, "Significant Events," p. 51.

21. "1985 Average Price of $33–43 for U_3O_8 Foreseen by DOE's John Patterson," *Nuclear Fuel*, October 31, 1977.

22. Some adjustment of prices to be paid under old contracts has been negotiated without involving litigation. For example, see "Pioneer Nuclear Got Uranium Price Relief It Sought from Philadelphia Electric," *Nucleonics Week*, February 26, 1976.

23. We refer to Rio Algom Mines as a U.S. corporation because it is incorporated in the United States, although its ownership is primarily British and Canadian. The U.S. antitrust laws apply to companies incorporated in the United States.

24. U.S. Congress, House, Committee on Interstate and Foreign Commerce, Subcommittee on Oversight and Investigations, *International Uranium Cartel, Hearings*, 95th Cong., 1st Sess., May 2, June 10, 16, 17, Aug. 15, 1977 (Washington: GPO, 1977).

25. 10 C.F.R. 40.23.

26. "Commonwealth Edison Reported to be First Utility Buyer of Foreign Uranium," *Nuclear Industry*, August 1970, pp. 45–46.

27. Byron E. Calame, "Uranium Cartel Members Often Busy Protecting Interests From Each Other," *Wall Street Journal*, June 23, 1977, p. 14.

28. "The uranium dilemma: Why prices mushroomed," *Business Week*, November 1, 1976, p. 97.

29. Nuclear Exchange Corp., "Significant Events," p. 50.

30. "Producers, Buyers Educate Each Other in Facts of Life," *Nuclear Industry*, April 1973, p. 54.

31. "Big Four Uranium Producers Weigh 'Orderly Marketing' Pact," *Nuclear Industry*, February 1972, p. 42.

32. U.S. Congress, House, *International Uranium Cartel*, p. 393.

33. Ibid., p. 448. The Justice Department investigated this meeting. See "Justice Department Uranium Probe Is Broadened," *Nuclear Industry*, July 1976.

34. U.S. Congress, House, *International Uranium Cartel*, p. 386.

35. Nuclear Exchange Corporation, "Significant Events," p. 49.

36. Gene Smith, "Uranium-Supply Lag Plagues Utilities," *New York Times*, November 12, 1973.

37. U.S. Congress, House, *International Uranium Cartel*, p. 408.

38. Nuclear Exchange Corporation, "Significant Events," p. 50.

39. U.S. Energy Research and Development Administration, *Survey of United States Uranium Marketing Activity*, ERDA 76-46 (Washington: USERDA, 1976), p. 6.

40. These figures refer to U_3O_8. Paul L. Joskow, "Commercial Impossibility, the Uranium Market and the Westinghouse Case," *Journal of Legal Studies* VI, 1 (January 1977):141.

41. "Renegotiating Fuel Pacts," *Nuclear Industry*, September 1975. See also Sanford L. Jacobs, "Utilities Rap Westinghouse Cancellation of Uranium Deliveries, Weigh Responses," *Wall Street Journal*, September 15, 1975, p. 8.

42. *United Nuclear Corp.* v. *General Atomic Co.*, "Trial Brief and Brief in Response to Defendant's Motion for Summary Judgment on the Issues of Anti-Trust; the Cartel; and Fraud," October 20, 1977, pp. 4, 35, 50.

43. Ibid., pp. 2, 61.

44. Ibid., p. 2.

45. Ibid., Attachment 1.

46. U.S. Dept. of Energy Survey of United States Uranium Marketing Activity, DOE/RA-0006 (Washington: U.S. DOE May 1978), pp. 2, 16, 23.

47. Joseph P. Mulholland and Douglas W. Webbink, *Concentration Levels and Trends in the Energy Sector of the U.S. Economy*, staff report to the Federal Trade Commission (Washington: GPO, 1974), p. 94.

48. 10 C.F.R. 60.5a. See also Arthur D. Little, Inc., *Competition in the Nuclear Power Supply Industry*, NYO-3853-1, TID UC-2 (Washington: USAEC, 1968), p. 160.

49. James K. Groves, "Uranium Revisited," *Rocky Mountain Mineral Law Institute* 13 (1967):95–96.

50. For the miners' viewpoint of this policy, see Raymond W. Taylor and Samuel W. Taylor, *Uranium Fever* (New York: Macmillan, 1970).

51. "Utah's Hidden Splendor Uranium Mine Has Resumed Operations," *Nucleonics Week*, April 15, 1976.

52. "An Option of the Zanzow Lease," *Nuclear Fuel*, November 1, 1976; "Lilco Has Purchased 'at least' 5-Million Pounds of Uranium," *Nucleonics Week*, February 12, 1976; "Intercontinental Energy Schedules Sale of Uranium," *Wall Street Journal*, December 7, 1976, p. 17.

53. USERDA, *Uranium Exploration Expenditures in 1975 and Plans for 1976–77*, GJO-103(76) (Grand Junction, Colo.: USERDA, 1976), p. 10.

54. Gulf Oil Corp., *1976 Annual Report and Form 10-K*, p. 17.

55. "GE Opens Ore-Buying Station as Hedge Against U_3O_8 Shortfalls," *Nuclear Industry*, August 1975; "GE Plans Second Ore-Buying Station," *Nuclear News*, December 1975; "Atlas Corp. Is Competing with GE for Uranium Ore in Utah and Colorado," *Nucleonics Week*, December 25, 1975.

56. John A. Patterson, "Uranium Supply Developments," paper presented at Atomic Industrial Forum Fuel Cycle Conference '76, Phoenix, Arizona, March 22, 1976, p. 3.

57. Patterson, "Status of Uranium Procurement," p. 6.

58. WPPSS Thinks It Has Set the Standards for Utility/Mining Company Contracts," *Nuclear Fuel*, October 31, 1977, pp. 5–7.

59. "Text of Fact Sheet on the President's Program Issued by White House Energy Staff," *New York Times*, April 21, 1977, pp. 48–49.

60. "Carter Proposes To Ban the Use of Plutonium," *Wall Street Journal*, April 8, 1977, p. 2.

61. Kerr-McGee Corp., *1975 Annual Report*.

62. Amendment to the "Agreement and Plan of Reorganization" in the GE-Utah merger. Reprinted in General Electric's "Notice of Special Meeting of Share Owners to be Held December 15, 1976" (Fairfield, Conn.: General Electric Co., 1976), Annex II.

63. Steven Rattner, "FTC Filing Suit Today to Block Arco from Acquiring Anaconda," *New York Times*, October 14, 1976.

64. "Bid to Bar Arco from Acquiring Anaconda Loses," *Wall Street Journal*, November 4, 1976, p. 6.

65. "Arco Acquires Anaconda Co. for $535 Million," *Wall Street Journal*, January 13, 1977, p. 9.

66. "Senate Kills Proposal Barring Oil Firms from Further Coal, Uranium Expansion," *Wall Street Journal*, September 9, 1977, p. 2.

8

The World Market

Since the price of uranium concentrate vastly exceeds the cost of shipping it, and since commercial U.S. imports and exports are now allowed by law, the domestic market is strongly influenced by the world market. In this respect, uranium is comparable to oil. By contrast, natural gas is not traded with other countries except Canada, since the cost of shipping it (as a gas) overseas is prohibitive. Liquefied natural gas can be imported from Algeria and elsewhere, but its cost is so high that it does not represent a major growth market even though it is part of the "deregulated" sector of domestic natural gas supply. Coal is traded from certain East Coast ports because shipping costs by freighter can come close to shipping costs by rail, and because metallurgical coal is scarce in Europe and Japan. However, only a fraction of domestic steam coal is affected.

Although uranium and oil are both sold in worldwide markets, some important distinctions should be noted. First of all, the domestic uranium resource base appears to be adequate to supply the lifetime requirements of all nuclear plants built by the year 2000, as we have seen in chapter 6. Although uranium prices may have to double in real terms to permit mining of resources in the $50 per pound "forward cost" category, such a price increase would not have a dramatic effect on the busbar cost of nuclear power. The oil situation is clearly different; coverage of the lifetime requirements of all oil-consuming machines built by the year 2000 is rarely even contemplated. Second, international trade in uranium concentrates will have an impact on the proliferation of countries producing nuclear explosives. Any peaceful nuclear explosion is a demonstration of a capability to produce nuclear weapons, and is evidence of the possession of nuclear weapons. Although oil trade has a financial impact on arms trade, there is no way that oil can be misused to develop or influence a vital weapons technology. Third, the uranium market is growing rapidly; projections of worldwide installed nuclear capacity in the year 2000 commonly involve a tenfold or greater increase over present capacity. Installed capacity in June 1977 was 94.841 GWe. At that time the OECD/NEA-IAEA estimate for the year 2000 was 1,000 to 1,900 GWe, excluding centrally planned economies.[1] A tenfold increase in oil consumption is inconceivable. Fourth, there is no publicly available evidence at present of a worldwide uranium cartel comparable to OPEC, although a cartel existed during 1972–1974. There are no "posted prices" for uranium.

In the following discussion, we shall examine the structure of the world uranium market, the conflicting political forces which determine foreign policy regarding the uranium market, the prospects for a scarcity of resources, the prospects for a cartel, and the current outlook for prices. To consider these topics with reference to the entire world rather than a handful of countries is an ambitious undertaking. However, there is no other way to present an accurate picture of the uranium market.

Market Structure

There are two ways to describe the structure of an international market: to refer to corporations or to refer to countries. We have prepared a partial list of uranium mining and milling corporations in Appendixes B and C. The corporations which purchase uranium concentrates are primarily utilities, although reactor vendors such as General Electric and Westinghouse also play an important role; information about both these groups is widely published.[2] However, we have chosen to base our discussion of market structure on countries rather than corporations, because political factors have such a strong influence on the uranium market, and because data on worldwide corporate reserves and production capabilities are relatively scarce.

A general summary of the market is presented in table 8–1, which is largely based on tables 8–2 through 8–5. Table 8–2 describes past production; table 8–3 describes reserves; table 8–4 describes production capacities; and they all refer to non-Communist countries. Table 8–5 gives us an idea of the relative scale and timing of various nuclear power programs. It is possible to get a rough idea of a country's need for uranium imports by comparing the 1985 capacity data in table 8–5 with the reserve data in table 8–3.

A country is likely to rely on net imports if its 1975 reserves are inadequate to supply the lifetime requirements of its 1985 nuclear capacity.[3] Each gigawatt of light-water reactor capacity will require (over 30 years) about 7,000 short tons U_3O_8 in ore, assuming no recycle, or about 5,000 tons assuming plutonium and uranium recycle (see chapter 4). The latter figure is not based on extensive empirical data, and in any event there is no point in trying to estimate these requirements precisely for our purposes because the capacity projections are uncertain. Heavy-water reactors require less U_3O_8 in ore than light-water reactors, but we doubt that the choice of reactor technology will determine whether or not a country becomes a net importer of uranium.[3] Sweden provides an exception to this rule of thumb regarding import status. Sweden's shale reserves have hardly been utilized.[4]

The various rows of table 8–1 distinguish the roles of different countries with regard to exports and imports. Under the heading Production Equal to Less Than 15 Percent of Consumption, we have grouped together countries which are totally reliant on imports and countries with modest production capacities attainable by 1985. The columns of table 8–1 distinguish the political and economic status of different countries in very general terms. Among developed, non-Communist countries, there exist a variety of viewpoints regarding the extent of safeguards which should be imposed on the commercial nuclear industry in order to diminish nuclear weapons proliferation. By *strict safeguards* we mean a policy of trying to prevent the construction of fuel reprocessing plants in countries other than the United States, United Kingdom, France, the Soviet Union, and the People's Republic of China, and of removing government support for reprocessing in the United States. In table 8–1 we have tried to group developed countries into two categories—proponents and opponents of such policies. This grouping does not do justice to the complexity of the issue, but permits important political distinctions to be made. Among developing, non-Communist countries, we distinguish those whose 1976 uranium production has been reported by the U.S. Bureau of Mines from those whose production has not been so reported.[5] The purpose of this distinction is to separate well-established producers from potential entrants into uranium supply.

Table 8–1 can be summarized by dividing the nations of the world into five groups.

1. *The uranium exporters.* Canada and South Africa are the leading exporters today, and Australia possesses large enough uranium reserves to become the third leading exporter. Significant quantities of uranium ore were first mined in Gabon in 1960, in Niger in 1970, and in South West Africa (Namibia) in 1976 (see table 8–2). The French company, Minatome, is exploring for uranium in Ireland,[6] and the International Atomic Energy Agency has sent British and Canadian exploration experts to Greece.[7]

In South America, the Brazilian government has joined the West German firm, Kraftwork Union, in a joint venture to explore for uranium in Brazil, subject to "guaranteed delivery of 20 percent of any ore to German utilities, with proportion increasing later."[8] Minatome is exploring in Colombia and French Guiana.[9] The IAEA sent uranium experts to Chile and Peru during 1977; moreover, "10 Latin American countries have received IAEA assistance in uranium exploration, mining and production of U_3O_8 concentrates" during 1972–1976.[10]

In Africa, many countries are attempting to become uranium suppliers. In 1975 Algeria, the Central African Republic (now the Central African Empire), and Zaire held reserves (see table 8–3). The Federation

Table 8–1
Probable Status of Various Countries with Respect to the Uranium Market in 1985

	I. Developed Countries in Favor of Strict Safeguards	II. Developed Countries Opposed to Strict Safeguards	III. Developing Countries with Reported U Production	IV. Developing Countries without Reported U Production	V. Communist Countries
Production for export only	Australia[a] Ireland[b]	Greece[a]	Niger Gabon S.W. Africa[e]	Algeria, Zaire, Central African E., Morocco, Malagasy R., Nigeria, Peru, Chile,[a] Columbia[b]	Czechoslovakia
Production with significant net export	Canada	South Africa		Brazil	
Production roughly equal to consumption	United States Denmark[c]	Portugal	Argentina	India Mexico Turkey	U.S.S.R. People's R. of China[f]
Production with significant net import	Sweden[d]	France Spain		Indonesia, Pakistan, Egypt, Philippines	Yugoslavia

Production equal to less than 15 percent of consumption	Finland	F.R. Germany, Japan, United Kingdom, Italy, Belgium, Switzerland, Netherlands, Austria, Luxembourg	Taiwan, South Korea, Iran	German Dem. R., Bulgaria, Romania, Hungary, Poland, Cuba[g]
Little or no production, with no nuclear power plants	Norway[a], New Zealand[a], Iceland		Kuwait,[b] Libya,[b] Saudi Arabia, Qatar, Iraq, Venezuela, Burma, Thailand,[b] Bangladesh,[a] Israel[b]	Albania, Vietnam, Laos, Mongolia, Cambodia, North Korea

[a] These countries have cancelled or indefinitely postponed plans to build specific nuclear power plants. Jamaica and Singapore have also cancelled their reactor plans. All these plans were in effect as recently as February 1974. See table 8–5 and U.S. Atomic Energy Commission, *Nuclear Power Growth, 1974–2000*, WASH-1139(74) (Washington: GPO, 1974) pp. 67–74.

[b] These countries plan to bring reactors into operation sometime after 1985. Hong Kong and New Caledonia also plan to operate reactors eventually. Such long-range plans should be regarded as tentative. See Atomic Industrial Forum, "Foreign Nuclear Electric Capacity Soars 33% in One Year, Now Surpasses U.S. Total," news release, June 10, 1977.

[c] Significant production from Greenland is assumed here. Denmark may cancel its nuclear program, following Norway's example.

[d] Sweden could eventually phase out uranium imports by mining its shale reserves.

[e] South West Africa is also known as Namibia. In 1975 the United Nations objected to South Africa's control over the area.

[f] There are no nuclear power plants operating in the People's Republic of China. Uranium is mined for military use.

[g] Cuba plans to build four nuclear power plants, but it has not stated whether any of them are planned to be in operation by 1985. See note b.

Table 8–2
Uranium Production in Non-Communist Countries, 1956–1976
(in short tons U_3O_8 in concentrate, except as noted)

Country	1956	1957	1958	1959	1960	1961	1962	1963	1964	1965	1966
I.											
Australia	300	400	700	1,100	1,300	1,400	1,300	1,200	370	370	330
Canada	2,280	6,636[d]	13,403	15,892	12,748	9,641	8,430	8,352	7,285	4,443	3,932
United States	6,000	8,640[d]	12,570[d]	16,420[d]	17,760[d]	17,399[d]	17,010	14,218	11,847	10,442	9,587
Sweden	6	10	10	10	10	10	10	10	10	20	50
Finland					40	20					
II.											
South Africa	4,365	5,700	6,245	6,445	6,409	5,468	5,024	4,532	4,445	2,942	3,286
Portugal						132	11	11	20	42	46
France		465	660	950	1,379[f]	1,619	1,978	1,987	1,331	1,421	1,542
Spain					60	55	55		77	67	66
F.R. of Germany				3	12	12	12				
III.											
Niger											
Gabon						428	514	582	586	724	616
Argentina	20	20	19	13	7	6	4	10	37	50	0
IV.											
R. of the Congo	1,300	1,300	2,300	2,300	1,200						
Malagasy R.		70	95	115		94	111	123[c]	196[c]	65[c]	65[s]
Fed. of Rhodesia and Nyasaland		25	50	38.							
Total	14,470[x]	23,270[y]	36,250[y]	43,350[z]	41,130[z]	36,312	34,512	31,025	26,204	20,586	19,520

Country	1967	1968	1969	1970	1971	1972	1973	1974	1975	1976
I.										
Australia	330	330	330	330						500
Canada	3,738	3,700	3,885	4,104	4,107	4,881	4,759	4,701	4,631	7,600
United States	9,125	12,368	11,609	12,905	12,273	12,900	13,235	11,528	11,600	13,500
Sweden	77	77	77	80	80	80	80	80		
Finland										
II.										
South Africa	3,360	3,882	3,979	4,119	4,189	4,000	3,411	3,389	3,096	3,200
Portugal	105	105	105	105	105	105	105	105		
France	592	1,626[c]	1,770[c]	1,944[i]	1,935[i]	1,940[i]	2,113	2,200	2,228	2,200
Spain	56	60	96	66	103	165	134	49		
F.R. of Germany										
III.										
Niger				42	474	958	1,046	1,232	1,820	2,000
Gabon	625	585[c]	536	416	601	577	712	852	1,209	1,200
Argentina	25	47	54	50	42	41	40	40		
IV.										
R. of the Congo	55[c]									
Malagasy R.		23[e]								
Fed. of Rhodesia and Nyasaland										
Total	19,098	22,803	22,411	24,161	23,909	25,647	25,635	24,176	25,034	30,650

Sources: U.S. Department of the Interior, Bureau of Mines, *Mineral Yearbook*, 1959 through 1974 (Washington: GPO, 1961–1976); and *Commodity Data Summaries 1977* (Washington: USDI, 1977), p. 183.

[c]Content of uranium ore produced.

[d]Deliveries to the Atomic Energy Commission.

[e]Content of uranothorianite exported.

[f]The figure for France includes the Malagasy Republic and Gabon.

[i]Produced in part from imported material. In 1970 the content of uranium ore produced in France was 1,744 tons U_3O_8.

[x]Includes an estimate for Italy, Japan, Morocco, Mozambique, Portugal, and Spain.

[y]Includes an estimate for Columbia, Italy, India, Japan, Morocco, Mozambique, and Portugal.

[z]Includes an estimate for Columbia, Italy, India, Japan, and Portugal.

Table 8–3
Reasonably Assured Resources of Uranium, 1977
(in thousand short tons U_3O_8; resources recoverable at costs no greater than $50/lb U_3O_8)

Country	Reasonably Assured Resources
I.	
Australia	384.8[a]
Canada	236.6
United States	835.9[b]
Sweden	391.3[c]
Denmark	7.5[d]
Finland	4.2
II.	
South Africa	452.4
Portugal	10.8
France	67.3
Spain	8.8[e]
F.R. of Germany	2.6
Japan	10.0
Italy	1.6
Austria	2.3
III.	
Niger	208.0
Gabon	26.0
Argentina	54.3
IV.	
Algeria	36.4
Central African Empire	10.4
Zaire	2.3
Somalia	8.1

of Rhodesia and Nyasaland (currently Rhodesia, Zambia, and Malawi) produced uranium from 1957 through 1959 (see table 8–2). Morocco has large amounts of uranium in phosphate deposits.[11] Italian and Japanese firms were exploring the Somali Republic in 1971, and Portuguese and West German firms were exploring Angola in 1972.[12] Exploration in the Malagasy Republic, Chad, Mauretania, Senegal, Guinea, Zambia, and Ethiopia was underway in 1975.[13] West German firms were exploring Nigeria and Togo in 1976.[14]

  2. *The uranium importers.* The net importers tend to fall into three groups. First, there are the Western European countries and Japan. The largest nuclear programs are in France, West Germany, Japan, and Spain (see table 8–5). The United Kingdom, Sweden, Italy, Belgium, Switzerland, Finland, the Netherlands, Austria, and Luxembourg have smaller nuclear programs in terms of 1985 capacity. Second, Taiwan, South Korea, and Iran are the developing countries with major nuclear power commitments and modest uranium reserves. South Korea reported re-

Table 8–3 continued

Country	Reasonably Assured Resources
Brazil	23.7
India	38.7
Mexico	6.1
Turkey	5.3
Philippines	0.4
South Korea	3.9
V.	
Yugoslavia	8.5
Total	2848.2

Source: This table is based on data found in OECD Nuclear Energy Agency and International Atomic Energy Agency, *Uranium: Resources, Production and Demand,* (Paris: OECD, 1977), p. 20. Figures have been converted from metric tons U to short tons U_3O_8 by multiplying by 1.3.

[a]A more recent estimate is 455,000 short tons U_3O_8. See A.J. Grey, "Current Australian uranium position," *Uranium Supply and Demand,* ed. by M.J. Spriggs and K.D. Casteel (London: Mining Journal Books, 1977), p. 212.

[b]This does not include 140,000 short tons U_3O_8 from byproduct production, 1975–2000. See U.S. Energy Research and Development Administration, *Statistical Data of the Uranium Industry,* GJO-100(77) (Grand Junction, Colo.: USERDA, 1977), p. 22.

[c]This includes 390,000 ST U_3O_8 in shale, the availability of which is severely restricted by environmental requirements. The concentration of uranium in this shale is 250–220 parts per million.

[d]This includes uranium in Greenland.

[e]This does not include lignite deposits, for which extraction costs are estimated to exceed $50/lb U_3O_8. In 1975 the amount of reasonably assured resources in lignite was estimated to be 105,040 short tons U_3O_8.

serves in 1977 (see table 8–3), and Iran is very active in uranium exploration, but publicly available information on Taiwan does not suggest that a substantial exploration program exists. Finally, Indonesia, Pakistan, Egypt, and the Philippines all have a modest amount of nuclear capacity planned for 1985 and exploration in progress. Domestic production would be preferable to imports for many reasons: energy self-sufficiency, a better balance of payments, increased employment, and possibly even the availability of uranium for nuclear explosives. It is difficult to tell whether or not these four countries will have uranium mines in operation by 1985, but we believe that the chances for success are good.

3. *The self-sufficient countries.* The danger of an interruption in supply, as a result of political disputes over nuclear safeguards or other issues, makes it very unlikely that a country with available uranium reserves adequate to fuel its nuclear industry will rely heavily upon imports.[15] Moreover, we suspect that countries with a modest "excess" of reserves over the lifetime requirements of nuclear reactors scheduled

Table 8–4
Uranium Production Capacities, 1977 and 1985
(*in short tons U_3O_8 per year; quantities which could be produced under optimal economic conditions*)

Country	Installed in 1977	Attainable in 1985
I.		
Australia	520	15,340
Canada	7,930	16,250
United States	17,550	31,000[a]
Sweden	—	220[b]
II.		
South Africa	8,710	16,250[c]
Portugal	111	351
France	2,860	4,810
Spain	248	1,654
F.R. of Germany	130	260
Japan	39	39
Italy	—	156
III.		
Niger	2,092	11,700
Gabon	1,040	1,560
Argentina	169	780
IV.		
Central African Empire	—	1,300
Brazil	—	501
India	260	260
Mexico	—	715
Turkey	—	130
V.		
Yugoslavia	—	234

Source: Except as noted, these data are from OECD Nuclear Energy Agency and International Atomic Energy Agency, *Uranium: Resources, Production and Demand,* (Paris: OECD, 1977), p. 24. Figures have been converted from metric tons U to short tons U_3O_8 by multiplying by 1.3.

[a]James Boyd, "United States Uranium Position," in *Uranium Supply and Demand* (1977), p. 146. The OECD/NEA-IAEA estimate is 46,800 ST U_3O_8/y.

[b]Owe Carlsson, "The Ranstad Project and Other Swedish Projects and Possibilities," in *Uranium Supply and Demand* (1977), p. 203.

[c]South African production is strongly influenced by the price of gold.

for operation by 1985 will tend to take a very cautious view of exports. Any depletion of domestic supplies will shorten the time period over which increases in nuclear capacity can be scheduled with confidence in fuel availability. While it is possible to adopt a wait-and-see attitude toward long-term nuclear power plans, many countries may decide to hold their uranium reserves in order to "firm up" these plans. In other words, potential exporters may withdraw from the export market. Denmark, Portugal, Argentina, Mexico, and India may be considered potential exporters, given the data in tables 8–3, 8–4, and 8–5, but we believe they will opt for self-sufficiency. The United States and Turkey did not have any "excess" reserves in 1975 to permit significant net exports.

In 1976 ERDA did "not anticipate exports to exceed 3 to 5 percent of industry's annual capacity between 1975 and 2000," but the General Accounting Office (GAO) suspected that 1975 exports were much larger. The GAO estimated that 20.4 percent of 1970–1974 production had been exported, but ERDA said that this figure was incorrect because the GAO worked with incomplete data. Recent DOE figures suggest that delivery commitments for import exceed commitments for export, but the trend in new contracts may reverse this pattern soon.[16]

Brazil might appear to be the exception; its reported reserves in 1975 were enough to fuel about 3 GWe, assuming recycle, over a 30-year period, yet it agreed to export uranium to West Germany. However, Brazil is not really risking a uranium shortage at all. The agreement with the Germans involves only the uranium which will be discovered as a result of the joint exploration program. The Brazilian share of this uranium may amount to a greater quantity than Brazil would have discovered if it had excluded the West Germans. This sort of exploration program may be adopted by the self-sufficient countries, excepting the United States.

4. *The Communist countries*. There is a possibility, of course, that additional countries will join the Communist bloc, but the current membership is easily identified. Although the Soviet Union has offered enrichment services to non-Communist countries, we know of no situation in which a Communist country has sold natural uranium in any form to a non-Communist country. Information about planned nuclear reactor capacity is not too difficult to obtain, but publicly available information on Communist uranium reserves and production is practically nonexistent.

5. *The nonnuclear, non-Communist countries*. There are four groups which can be distinguished within this category. First, there are three developed countries: Norway, New Zealand, and Iceland. Norway and New Zealand have cancelled their nuclear power plans, perhaps because hydroelectric power is much cheaper for them. Iceland is probably too small to purchase power plants in the size range where nuclear becomes competitive with coal.[17] Uranium exploration in these countries is possible, since Ireland and Greece have set an example, but controversies similar to the Australian debate on mining would probably result from such activities.

Second, there are the OPEC oil producers—Saudi Arabia, Kuwait, Iraq, Abu Dhabi, Qatar, Libya, and Venezuela. It is curious that OPEC countries have adopted very different policies with regard to nuclear

Table 8–5
Nuclear Power Capacities
(*in net gigawatts*)

Country	Installed in 1977[a]	AIF Projection for 1985[a]	SIPRI Projection for 1984[c]
I.			
Canada	4.000	12.600	10.261
United States	47.186		151.744[b]
Sweden	3.760	9.500	9.506
Denmark	—	0.900	
Finland	0.420	3.200	2.160
II.			
South Africa	—	1.844	1.850
Portugal	—	0.650	
France	6.535	50.000	29.418
Spain	1.120	19.675	15.091
F.R. of Germany	9.500	30.000	35.116
Japan	7.428	49.000	21.368
United Kingdom	6.900	11.000	10.710
Italy	1.400	?	5.278
Belgium	1.387	.137	6.493
Switzerland	1.006	3.801	6.833
Netherlands	0.505	3.505	0.499
Austria	—	2.000	0.692
Luxembourg	—	1.300	
III.			
Argentina	0.345	1.545	0.919
IV.			
Brazil	—	1.926	3.116
India	0.800	?	1.690
Mexico	—	1.308	1.308
Turkey	—	0.600	
Iran	—	2.400	4.200

power. Iran has made a major commitment to nuclear capacity, Indonesia has more modest plans, Algeria has uranium reserves for export, and Nigeria is being explored, Kuwait and Libya have vague plans to build reactors eventually, and the rest of OPEC appears to have no plans for either reactors or uranium mines.

Third, there are some countries outside OPEC with low per capita incomes: Ethiopia, Thailand, Burma, and Bangladesh, for example. Thailand still plans to have a nuclear reactor operating by 1990; Bangladesh announced a reactor at Roopur and then cancelled the plans. Nuclear power plants are of little use in these countries unless extensive transmission and distribution systems are built. The type of commercial nuclear plants on order today are probably too large to fit into the transmission systems of many developing countries. According to I.H. Usmani,

Table 8–5 continued

Country	Installed in 1977[a]	AIF Projection for 1985[a]	SIPRI Projection for 1984[c]
Indonesia	—	1.600	
Pakistan	0.125	1.325	0.726
Egypt	—	1.200	
Philippines	—	1.226	0.600
Taiwan	0.636	5.144	4.922
South Korea	0.595	3.724	1.798
Thailand	—	—	0.600[d]
V.			
Czechoslovakia	0.552	3.872	3.391
U.S.S.R.	7.905	31.175	21.816
Yugoslavia	—	1.532	1.432
German Dem. R.	1.400	?	4.959
Bulgaria	0.880	1.760	1.701
Romania	—	1.640	0.440
Hungary	—	?	1.224
Poland	—	0.880	
Cuba	—	?	

[a]Atomic Industrial Forum, "Foreign Nuclear Electric Capacity Soars 33% in One Year, Now Surpasses U.S. Total," news release, June 10, 1977.

[b]The Department of Energy projection is 127 GWe installed in the U.S. by 1985. See R.W. Bown and R.H. Williamson, "Domestic Uranium Requirements," *Uranium Industry Seminar*, GJO-108 (77) (Grand Junction, Colo.: USDOE, 1977).

[c]Stockholm International Peace Research Institute, *World Armaments and Disarmament: SIPRI Yearbook 1977* (Cambridge, Mass.: M.I.T. Press, 1977), pp. 38–39. These figures are based on data from the International Atomic Energy Agency, from *Nuclear Engineering International* 21: 238–251 (January-December 1976), and from *Free China Weekly*, May 9, 1976.

[d]The Atomic Industrial Forum estimates that Thailand will have 600 MWe by 1990, but not by 1985.

head of Pakistan's Atomic Energy Commission from 1960 to 1971, using nuclear power in the rural areas of developing countries would be like "using a 12-gauge shotgun to kill a fly."[18] Moreover, the capital intensity of nuclear power is a major economic disadvantage in these areas. The political stability of the governments involved may be inadequate to attract foreign investment in uranium mining and milling.

Fourth, there are developing countries outside OPEC which do not appear to have sufficiently large urban populations to justify nuclear power plants on economic grounds. Israel and Jamaica are two examples. Of course, Israel is unique in many respects and may be postponing commercial nuclear plants in order to avoid IAEA inspection of its other facilities, or in order to avoid reliance upon a facility which is an easy target for attack.

On the whole, the market appears to be moving toward a situation in which countries which have just enough reserves to meet their own requirements will withdraw from international trade in uranium. This leaves non-Communist importers and non-Communist exporters; in short, Australia and Canada are competing with South America and Africa for sales to Western Europe, Japan, Taiwan, South Korea, and Iran.

The Dilemma of Nuclear Proliferation

There are three methods by which uranium concentrates can be used to manufacture the essential materials required for nuclear explosives. First, the uranium may be placed in a reactor which is designed simply to convert the U-238 isotope into plutonium-239. After exposure within the reactor, the fuel must be chemically processed to isolate the plutonium, which then can be used in explosives. Second, the U-235 isotope may be separated from U-238 by one of several technologies. Gaseous diffusion is a method of isotope separation involved in the existing U.S. enrichment plants, and centrifuge enrichment is a method to be used in the additions to U.S. capacity which are now under construction. U-235 can be used in explosives when it is in a relatively pure form. Third, the uranium may simply be used to fuel commercial nuclear reactors, and plutonium will be produced inevitably as a byproduct of electricity production, since all commercial fuels contain U-238. This plutonium can be recovered by fuel reprocessing.

If plutonium had no commercial value, it would be relatively easy to restrict nuclear proliferation by preventing the construction of fuel reprocessing plants. However, plutonium can substitute for U-235 as a nuclear fuel. If nuclear fuel could be left in a reactor until nearly all the U-238 had turned to plutonium-239 and nearly all the plutonium-239 had been fissioned, there would be no need for reprocessing. However, the fissioning of uranium and plutonium creates waste products which "poison" the chain reaction.

The present policy of the United States is to discourage commercial fuel reprocessing in order to increase the difficulty of obtaining plutonium for use in nuclear weapons. Since plutonium is a substitute for U-235, a shortage of uranium concentrates would increase the economic attractiveness of recovering plutonium. Conversely, one of the simplest ways to discourage reprocessing is to make uranium concentrates as cheap and plentiful as possible. Unfortunately, there are several difficulties with the "cheap uranium" policy; these can be illustrated by considering the present situation facing four countries.

Australia's Dilemma

Australia has very large uranium reserves and no domestic nuclear power plants on order. One might suppose, therefore, that the government would be in favor of exporting as much uranium as possible in order to diminish, if not eliminate, commercial plutonium use. This policy would not only inhibit nuclear proliferation, but create jobs through the expansion of the mining industry. No domestic reactor would be in danger of running out of fuel, and no domestic reprocessing plant would be driven out of business. However, the actual political situation in Australia does not demonstrate support for such a policy.[19] There are several reasons why various groups might oppose it.

1. In order to maximize sales, Australia would have to minimize profits, that is, sell uranium as cheaply as possible without driving the mining companies out of business. This is obviously not in Australia's financial self-interest; either the mining companies would suffer, or the taxpayers would have to subsidize them.

2. By driving down the price of uranium, Australia would threaten to put all commercial reprocessing companies out of business, regardless of nationality. The governments of Great Britain, West Germany, and France have already committed themselves to reprocessing, and even though their utilities would be pleased to have cheap uranium, the governments would be humiliated. Demonstration of reprocessing technology may be a matter of national pride, and investments have already been made in it. Australia might not want to strain its relations with these countries, even if nuclear proliferation is at stake.

3. A plentiful supply of uranium would tend to increase the economic attractiveness of nuclear power and thereby increase the amount of plutonium produced as a byproduct. Conversely, a scarcity of uranium would diminish nuclear capacity growth. Although a large nuclear industry with no reprocessing is less hazardous, with regard to proliferation, than a small nuclear industry with reprocessing, the idea that a choice must be made between these two alternatives is politically unpopular. Moreover, the safety and health problems associated with nuclear power plants may be considered more important, in the eyes of Australian voters, than the safety, health, and proliferation problems associated with reprocessing and plutonium fuel fabrication. For example, nuclear waste will exist with or without reprocessing, and it may be argued that the best policy is to create as little of this waste as possible.

4. It may be argued that the most ethical way to deal with nuclear proliferation is to refuse to sell uranium except under strict safeguards agreements, or perhaps even to refuse to sell it at all. These conditions

would be imposed in order to ensure that no Australian uranium would be used to make weapons, regardless of whether or not other sources of uranium were used to make them. In other words, Australians might view their ethical responsibility in the uranium market simply in terms of their own uranium resources. This is not much consolation to the rest of the world, since it strengthens the economic incentives to reprocess fuel, but it may rid Australians of a guilty conscience.

5. Finally, Australians might speculate that the efforts of the United States, Canada, Sweden, and others will suffice to halt fuel reprocessing in all countries except those which have no interest in its commercial profitability. For these intransigent reprocessors, it does not matter whether uranium concentrates are cheap or expensive, plentiful or scarce; the plutonium will be recovered simply for military reasons. Under these conditions, Australia might as well maximize its profits and royalties from uranium mining.

In short, there are several reasons why a "cheap uranium" policy in Australia would be politically unpopular. We have called this situation Australia's dilemma, but it is a dilemma faced by every country which is deeply concerned about proliferation and holds an excess of uranium resources over its own requirements. Canada, for example, faces the same issues.

The United States' Dilemma

In 1973 the U.S. government launched a National Uranium Resource Evaluation program to provide an assessment of potential resources and to identify favorable areas for industry exploration (see chapter 2). Although this program was originally intended to support a growth in nuclear capacity well above current projections, the information on potential resources has become useful in assessing the need for domestic reprocessing. As we saw in chapter 6, there appear to be enough uranium deposits to provide fuel for the lifetime requirements of reactors operating in the year 2000 without reprocessing. Although demonstrating the geological availability of uranium does not guarantee fuel availability, it is the first step, and probably the most important one. The United States could try to diminish worldwide concern over a long-term uranium shortage by making a major commitment of funds to the International Uranium Resource Evaluation. The United States could subsidize uranium exploration in all non-Communist countries in order to expand their reserves and prepare some estimate of their combined potential resources. Although the International Fuel Cycle Evaluation may move in this direction, the United States has not adopted a "cheap uranium" policy. The domestic popularity of such a policy would be diminished by five major considerations.

1. The domestic uranium mining industry would demand protection against low-cost imported uranium if a worldwide surplus of uranium developed. If protection were given, relations with Canada and other uranium-producing countries could be strained. The restrictions on enrichment of imported uranium for use in domestic reactors would have to be reimposed if the viability of the domestic mining industry were threatened. These restrictions have been a source of contention since 1964, when they were established by the Private Ownership of Special Nuclear Materials Act. Another statute which would require protection of the domestic uranium industry is the Energy Reorganization Act of 1974, since it requires the Department of Energy to promote energy independence. Reliance on imported uranium would be contrary to the objectives of this statute.

2. As noted earlier, a plentiful supply of uranium would increase the economic incentive for countries to "go nuclear." A substantial portion of the American public is opposed to nuclear power in any form, as a result of concerns about health and safety. These individuals would prefer to restrain, if not eliminate, the growth of world nuclear capacity rather than try to make the nuclear industry safer from a proliferation standpoint.

3. Provision of assistance to foreign countries through uranium exploration is a "carrot" rather than a "stick." It may be argued that if a country is interested in constructing a reprocessing plant, thereby making it easier to divert plutonium from commercial use into military use, that country should be punished rather than rewarded. In other words, the objection may be raised that "carrots" are unethical. A comparable sort of claim would be that heroin addicts should be punished rather than provided with free medical care. The IAEA is sending experts to Brazil and Pakistan to explore for uranium, although these countries plan to build reprocessing plants. The United States and the Soviet Union "pay a large share of the IAEA budget."[20] Such uses of U.S. funds eventually may become a focus of controversy.

4. The cost of the International Uranium Resource Evaluation may be seen as an extravagant use of the taxpayer's money—a form of assistance which neither fulfills essential human needs nor guarantees an improvement in U.S. security. The Congress may be reluctant to appropriate the funds or to provide tax incentives for U.S. corporations engaged in foreign uranium exploration programs.

5. As a major military and economic power, the United States has a

variety of means at its disposal to discourage nations from reprocessing spent fuel. It may be argued that some of these are likely to be just as effective as a "cheap uranium" policy, and yet free of the disadvantages.

Sweden's Dilemma

In 1975 Sweden possessed larger uranium reserves than any other non-Communist country except the United States, yet it had no production capacity (see tables 8–3 and 8–4). Sweden is an importer, but it could be an exporter. Since Swedes are concerned about nuclear proliferation, their behavior may seem paradoxical; they are contributing to a scarcity of uranium by withholding one-sixth of the non-Communist reserves from the market. Energy independence lies within their reach, but they are in no hurry to achieve it.

The principal reason for Swedish reluctance to mine uranium appears to be a concern for environmental quality. Despite elaborate plans to minimize disturbances to the environment, the Ranstad 75 project was cancelled after the 1976 elections, in which the Social Democratic party lost power. This project would have involved the mining of 6 million metric tons per year of shale. At present a smaller project is envisioned, involving 1 million metric tons per year.[21]

To some extent, all developed countries with uranium reserves face a tradeoff between nuclear proliferation hazards and the environmental impact of domestic uranium mining. Sweden's case may be extreme, but the United States, Canada, and Australia are also experiencing controversies over the hazards of mill tailings and the need for reclamation of surface-mined land. Restrictions on mining tend to raise uranium prices and reduce uranium production, thereby strengthening the economic case for reprocessing. It may be argued that environmental regulations provide a tangible and certain benefit, whereas reductions in the dollar value of plutonium oxide yield only a nebulous and uncertain reduction in risk. On the other hand, nuclear proliferation may be considered a vastly more important issue than tailings management and land reclamation.

Britain's Dilemma

Although Great Britain does not oppose the creation of a domestic commercial reprocessing industry based on facilities formerly devoted to military use, Britain was the first nuclear weapons state to ratify the Nuclear Non-Proliferation Treaty.[22] Unlike France and West Germany, Britain does not hold a contract to build a reprocessing plant in a foreign country.[23] Thus the British seem to favor a "double standard" for reprocessing technology: it is acceptable in nuclear weapons states and in Western Europe, but unacceptable in developing countries. A competitive uranium market, however, does not observe such a double standard. A plentiful supply of uranium would not only cut into the profits (or increase the losses) of Brazilian reprocessors, it could also drive British Nuclear Fuels Limited out of the reprocessing business.

There is a lot of domestic political opposition to reprocessing in Britain, not only in relation to the proliferation issue but also in relation to health, safety, and environmental issues. It would not be too speculative to suppose that sometime in the next few years the government might adopt a position similar to the current official policy of the United States. This change would then clear the way for a "cheap uranium" policy, which would further complicate the embarrassing and frustrating issue of South African uranium.

From 1968 through 1971, South Africa produced more uranium concentrate than any other non-Communist country except the United States. Since 1972 Canada has taken second place, but South Africa may again surpass Canada by 1985 (see tables 8–2 and 8–4). If we accept the estimate of 16,535 short tons U_3O_8 produced in South Africa in 1985,[24] and assume a price of $40 per pound in 1977 dollars, annual uranium production would be worth $1.32 billion. If we assume further that South Africa has 1.844 GWe installed by 1983,[25] requiring 200 short tons U_3O_8 per gigawatt per year, export revenues in 1985 would be about $1.29 billion in 1977 dollars.

No restriction on purchases of South African uranium can be imposed or recommended by Great Britain without increasing the economic incentives for plutonium recycle. An embargo, for example, would clearly create a shortage of worldwide uranium supplies. Economic pressure cannot be put on South Africa, through restrictions on uranium exports, without increasing the hazards of nuclear proliferation. In fact, South Africa itself might develop nuclear weapons in response to military threats from the rest of the world.

Rio Tinto Zinc (RTZ), a British mining company, announced plans in 1968 for construction of a mine in the Territory of South West Africa (Namibia), which is controlled by the Republic of South Africa. The United Nations declared South Africa's control of the area illegal, but Rio Tinto Zinc went ahead with construction, after a 2-year delay. The British Atomic Energy Authority contracted for a portion of the uranium. The situation in August 1975 was described as follows:

The British Labour government is acutely embarrassed by the fact that its public stands on Africa and Namibia conflict with the British need—

and determination—to continue obtaining uranium through RTZ from the Rossing mine.[26]

A year later, the problem was no less complex:

> This mine is scheduled to supply Japan, Germany, and Britain. Rio Tinto has begun quiet contacts with the South West African People's Organization (SWAPO) just in case their faction wins the argument over the future of this territory.[27]

The British role in South West Africa provides an example of a general difficulty regarding uranium producers. If the developed countries try to maximize world uranium production, they will have to provide economic support to regimes which are not democratically governed. The Central African Empire holds uranium reserves (see table 8–3). The Federation of Rhodesia and Nyasaland produced uranium during the 1950s; perhaps Rhodesia will enter the market in the next decade. As noted earlier, Chile is involved in uranium exploration. The conflict between uranium supply and democratic values is ultimately not just Britain's dilemma, but everyone's dilemma.

The topic of nuclear proliferation involves a lot more, of course, than removing the economic incentives for plutonium recycle. We have focused on a narrowly defined topic. However, it does appear that the policy of expanding world uranium supplies, particularly in Africa and South America, deserves more attention than it has received to date.

Prospects for a Uranium Shortage

The latest resource estimates prepared by the OECD Nuclear Energy Agency and the International Atomic Energy Agency include 2.2 million metric tons of uranium in reserves and 2.1 million metric tons of uranium in probable resources, at costs up to $50 per pound U_3O_8.[28] These figures do not include uranium in Communist countries, with the exception of 27,000 metric tons of uranium in Yugoslavia. Assuming lifetime requirements of about 7,000 short tons U_3O_8 in ore per gigawatt of nuclear capacity, non-Communist resources are sufficient for 794 GWe. With plutonium recycle, of course, an even greater capacity would be covered. According to figures released by the Atomic Industrial Forum, the United States had 199 GWe operable, under construction, or on order as of September 14, 1978, while foreign countries had 222 GWe of such capacity as of June 1978 (see table 8–6). Therefore, an additional 373 GWe are covered by non-Communist uranium resources, and some further capacity is covered by Communist resources. Given the recent progress of

Table 8–6
Nuclear Power Commitments Outside the United States, 1974–1978
(*in net gigawatts*)

Category of Nuclear Capacity	1974	1975	1976	1977	1978
Operable	24.293	29.175	35.773	47.655	56.350
Under construction	50.097	59.767	85.182	90.943	126.618
On order	56.112	54.462	53.787	42.135	38.744
Subtotal	130.502	143.404	174.742	180.733[a]	221.712
Planned	90.073	150.874	168.504	182.058	214.300
Total	220.575	294.278	343.246	362.791	436.012

Source: Atomic Industrial Forum, "Nuclear Power-Plant Commitments Outside the U.S. Climb 17% in Year," news release, June 2, 1976; "Foreign Nuclear Electric Capacity Soars 33% in One Year, Now Surpasses U.S. Total," news release, June 10, 1977; "Non-U.S. Nuclear Energy Commitments Up 20% in a Year; 52 Countries List Reactor Programs in Latest AIF Survey," news release, August 31, 1978.
[a]This includes 152.554 GWe in non-Communist countries and 28.179 GWe in Communist countries.

nuclear power in Europe and Japan, it does not seem likely that these 373 GWe will be installed by the year 2000, if ever. In short, unless the OECD/NEA and IAEA have overestimated probable resources, there seems to be no danger of a scarcity of uranium resources in the total non-Communist world.

Official projections do not provide an accurate picture of foreign nuclear programs. On October 5, 1977, "the Common Market energy Commissioner, Guido Brunner, told the I.E.A. meeting that the Common Market countries would now be lucky to produce 90,000 megawatts of nuclear power in 1985, instead of the 160,000 megawatts it is officially planning to have."[29]

In the light of these figures, it is pertinent to ask how any country could seriously worry about a shortage of uranium. While it may be argued that presently known resources will not support nuclear capacity growth beyond the year 2000 without a sharp reduction in the growth rate, it is unrealistic to believe that governments or utility companies feel threatened today by the possibility of a shortage 25 years from now; there are much more urgent problems to worry about. Why, then, are countries in such a hurry to build breeder reactors and reprocessing plants? Several reasons may be given.

1. In developed countries, the primary motive for reprocessing, in our view, is energy self-sufficiency. Great Britain, France, and West Germany will be able to reduce their dependence on uranium imports by recycling plutonium. The German utilities accuse the United States of

using uranium as a political weapon in connection with safeguards policies:

> Any misuse of uranium supplies as a political weapon must cease as soon as possible.
>
> . . . Nuclear opponents can quite successfully use the argument that uranium supplies are even less secure than are oil imports, because of political factors.[30]

This statement is directed especially toward the United States and Canada. The United States has withheld enriched uranium in connection with its nuclear proliferation policy, and European enrichment capacity has been planned to develop greater self-sufficiency.

2. In developing countries, a major concern is likely to be the acquisition of a capability to produce nuclear weapons. Although there may be no urgent plan to produce nuclear weapons, countries may feel more secure if the technology for obtaining plutonium is available when needed. It is also possible, although improbable in our view, that reprocessing technology may be given to a developing country (with some token payment to suggest a sale) by a supplier who believes it strategically imprudent to construct the facility on its own soil.

Of course, it may be argued that energy self-sufficiency is a primary concern, but if this were true, any decision to import a capital-intensive reprocessing plant would be premature until a thorough exploration program had demonstrated the absence of domestic uranium resources. Note that it is in the interest of a country seeking a nuclear weapons capability to vastly overstate its nuclear capacity growth, in order to pretend that its uranium requirements will not be as modest as they actually will be. Pakistan is a case in point.[31]

3. The availability of Swedish shale and Spanish lignite reserves are uncertain, primarily because of the environmental problems involved in mining them. Domestic political issues as well as technical factors tend to constrain production from these sources.

4. The production rate of uranium from mines in which it is recovered as a byproduct will depend on the market price and outlook for the other product or products. All uranium mined in South Africa is a byproduct of gold mining, with the exception of the Palabora copper mine. The sharp reduction in uranium output in 1973 (see table 8–2) demonstrates a response to gold price increases; lower-grade ores suddenly became worth mining. The Cluff Lake deposit in Canada and the Jabiluka Two deposit in Australia are both to be mined for gold as well as uranium. Present plans for Swedish shales involve the recovery of some or all of the following: aluminum, potassium, magnesium, phosphorus, sulfur, vanadium,

molybdenum, nickel, and heat from the combustion of organic matter.[32] United States reserves include 140,000 short tons U_3O_8 recoverable as a byproduct of phosphate or copper mining.[33] Morocco may also recover uranium from phosphate mines.

5. The availability of Australian uranium is subject to domestic political debate over a number of issues—nuclear proliferation, nuclear wastes, the hazards of uranium mining and milling, land reclamation, the rights of Aborigines, the creation of national parks, and foreign ownership and control of Australian corporations. The long-term outcome of this debate is, of course, difficult to predict.

6. Delays in the construction and operation of new enrichment plants are possible. A shortage of enrichment capacity would require the operating tails assay to be increased, thereby raising uranium requirements.

7. Finally, some governments may believe that non-Communist nuclear capacity will be much greater than 794 GWe by the year 2000.[34] Although these expectations are inconsistent with recent trends, the nuclear industry may hope for a dramatic reversal.

The countries which face the greatest risk of a uranium shortage are the importers—Western Europe, Japan, Taiwan, South Korea, and Iran. In a sense, their view of uranium resources can become a self-fulfilling prophecy. If they believe that it is better to invest money in reprocessing technology rather than uranium exploration, because the latter is likely to be unsuccessful, the rate of growth of world reserves will fall below the level which would result from the opposite decision. This low growth rate may then be regarded (incorrectly) as evidence of uranium scarcity.

One of the major uncertainties in the overall outlook is the role of Communist countries. They might enter the market as either exporters or importers. However, we know of no evidence that either course of action is likely.[35]

Prospects for a Uranium Cartel

The success of OPEC has made it clear that no analysis of an international market is complete without an examination of the prospects for a cartel. There are two basic types of cartels—those among corporations and those among governments—but hybrid forms are also possible. We shall address the subject by stating the conditions favorable for cartelization, in theory, and examining the relevance of each condition to the uranium market.[36] For present purposes, it is best to think of the worldwide uranium supply industry in terms of three sectors: production in the

Communist countries (excluding Yugoslavia), in the United States, and in the rest of the world. The first sector is not involved in trade with the other two, and we shall ignore it. The Communist sector is probably characterized by monopoly control, that is, Soviet control over all production except Chinese production, which probably represents a small portion of the total. In the second sector, corporations are excluded from cartel participation by the U.S. antitrust laws. The third sector is partially restrained by antitrust laws. For example, under U.S. law, domestic corporations may not allow their foreign subsidiaries to fix prices unless it can be shown either that such actions do not affect the prices of goods purchased in the United States or that such actions were dictated by foreign governments and could not have been avoided by subsidiaries of U.S. corporations (except by going out of the uranium business).

The following discussion is based on the economic incentives for cartel formation in the non-Communist, non-U.S. sector. We will not examine the legal restraints on such cartel activity, but we do not consider this omission to be a serious one. Participation by subsidiaries of U.S. corporations can be protected by explicit orders given by foreign governments. The actual existence of a uranium cartel between 1972 and 1974 (see chapter 7) demonstrates that the laws of Canada, Australia, Great Britain, France, and South Africa have been ineffective in restraining price-fixing activity. The West German antitrust laws appear to have successfully kept Uranerz and Urangesellschaft out of the cartel, but these two companies are unlikely to "break" a cartel by selling competitively priced uranium to utilities outside West Germany. Even if all uranium-mining corporations refrained from cartel activity, it would still be possible for the governments of producing companies to nationalize the mines and mills and then cartelize, or to require payment of royalties and severance taxes which would have the same effect as price-fixing agreements.

The following conditions provide a favorable setting for cartel formation:

1. *Separate identity of producers and consumers.* In the case of corporations, this involves an absence of vertical integration—the less there is, the easier it will be to form a cartel. The mining subsidiaries of utilities, for example, are not going to conspire to increase the price of uranium delivered to these same utilities. In the case of countries, a cartel is unlikely unless some nations consume more than they produce. By definition, this means that other nations must produce more than they consume. If some producers do not consume the product at all—for example, if uranium-exporting African nations do not build nuclear power plants—conditions for cartelization are relatively favorable. Those nations which produce exactly as much as they consume will not inhibit

price fixing; they will simply become self-sufficient and withdraw from international trade in the product in question.

In the uranium market there is a trend toward vertical integration. If a national government owns an electric utility and a portion of a mining company which supplies that utility, we may consider the utility to be vertically integrated. If a corporate cartel were formed, utilities could "get around" it by forming their own mining ventures. Therefore, a cartel among nations seems more likely than one among corporations. There are, of course, producers such as Australia, Niger, and Gabon, with no power reactors planned.

2. *Inelastic demand, especially in conjunction with elastic supply.* This condition is illustrated in the upper left corner of figure 8–1. The advantage to a cartel of inelastic demand is obvious: price increases will yield revenue increases. Like a monopolist, a cartel will raise prices up to the point where marginal revenue equals or exceeds marginal cost. The supply curves in figure 8–1 are all marginal cost curves; that is, we have assumed in these examples that marginal cost always equals or exceeds average cost. Long-run supply curves, which are based on new production facilities, are always more elastic than short-run supply curves, which are based on existing facilities. Although an elastic supply curve makes it easier for one member of a cartel to "cheat" on the other members, there is little or no incentive to cheat if market shares under the cartel are similar to the market shares which would exist under competition (see below). The disadvantage of an elastic supply curve—easy cheating—is outweighed, in our opinion, by its "punitive" price level; if the cartel members do not act in unison, they will be "punished" by a much greater drop in price than would result from an inelastic supply curve (see figure 8–1). The low price levels prevailing under competition reduce the incentives for a member to cheat on the cartel.

The uranium market, judged by this criterion, is well suited to cartelization. Utility demand for uranium is rather inelastic (see chapter 4), and long-run supply, through investment in exploration, mines, and mills, is elastic. One of the factors which is difficult to estimate, however, is the ability of utilities to reduce their uranium requirements through fuel reprocessing and plutonium recycle. The lead time required to "speed up" a reprocessing industry, in response to a uranium cartel, is uncertain. The costs of reprocessing and recycle, and therefore the uranium price at which these activities become profitable, are uncertain. The lead time required to expand enrichment capacity, and thereby reduce the operating tails assay, is probably at least 8 years. The lead time for developing a large-scale breeder reactor industry, in response to a jump in uranium prices, is so long that in our view breeders are not a form of retaliation which a cartel need fear.

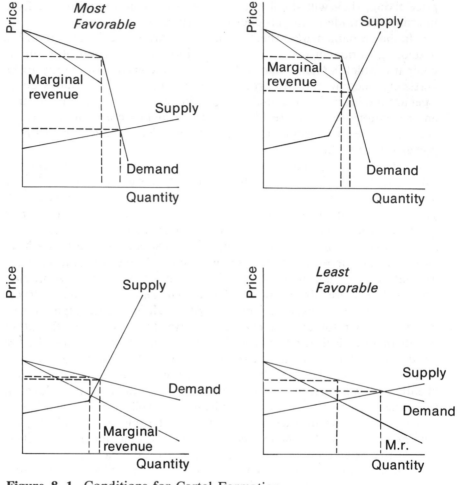

Figure 8–1. Conditions for Cartel Formation.

3. *Cartel market shares similar to competitive market shares.* The producers which form a cartel must agree upon a method of allocating shares of total production at the cartel price level. If one producer is allotted a much smaller share than he would be able to achieve in a competitive market, then he has an incentive to "cheat" on the cartel, that is, cut prices in order to expand the volume of his sales. Consequently, in the interest of unity, it is best for the cartel to cut back each producer's "competitive" production level by an equal percentage. A serious difficulty arises when the cartel price is high enough for new

producers to enter the market—producers who would be driven out of business by competitive pricing. For example, OPEC would not want to raise oil prices up to the level where oil shale, coal gasification, and alternate energy sources could rapidly replace oil use. Although it is possible to admit the "new" producers into the cartel, their share of production simply cuts into the shares of "old" producers and reduces or even eliminates the incentives for the latter to fix prices.

In this context, conditions in the uranium market are very unfavorable for cartel formation. Uranium price increases are likely to bring a large number of new producers into the market. Current exploration in Africa and South America demonstrates that a lot of countries are at least trying to enter the market. Most of the mining subsidiaries of utilities can also be regarded as new producers.

4. *A fungible product with low shipping costs*. Price fixing is simplest when the product is of uniform quality, or at least of a quality which can easily be measured by an objective standard. Moreover, it is relatively easy to fix the price of a product which is marketed competitively at prices which vary only slightly from one point of delivery to another rather than a product which is sold at substantially different prices in different cities. Geographically uniform prices are achieved when shipping costs are low relative to production costs—in other words, when the product is not bulky.

From this standpoint, uranium concentrate is an ideal commodity for price fixing. The value of this concentrate is measured by its U_3O_8 content; at \$40 per pound U_3O_8, shipping costs are negligible. Uranium ore, of course, has just the opposite qualities, and is unlikely to fall under a cartel's direct control.

5. *Rising marginal costs*. If marginal costs rise as the level of production increases, market shares in a competitive situation will stabilize at the values which equate each producer's marginal cost to the price. These stable market shares are needed to establish a basis for the cartel's production quotas. By contrast, if marginal costs declined as production increased, there would be a "natural monopoly" situation—the largest producers would force the small ones out of business and compete or merge until only one producer remained. Like most products, uranium concentrate is subject to rising marginal costs, and is therefore suited to cartelization.

6. *Small number of producers*. It is easier to negotiate a cartel agreement if the number of producers is small. Since the producers who could not stay in business at competitive prices will probably be excluded from the cartel, their number is relatively unimportant. The number of producers in a competitive market is the crucial figure.

According to this criterion, conditions in the uranium market are unfavorable for cartel action. As we have noted earlier, there are a lot of new producers trying to enter the market now, and higher prices would bring forth still more producers.

7. *Large number of consumers.* If a producers' cartel is established, it is easier for consumers to negotiate an agreement about methods of retaliation if the number of consumers is small. For example, if demand for the product can be reduced by a research and development program or a manufacturing facility which has a production capacity larger than any single consumer's requirements, the consumers may pool their resources. In many situations, however, there may be no form of retaliation which can be imposed, or else there may be methods which do not require collective action.

If a cartel tries to discriminate among consumers and behave like a perfectly discriminating monopolist, it is in the collective interest of consumers to band together and agree that whoever is offered the lowest price will sell as much as possible to the rest. However, it may be in the individual interest of the low-price consumer to resell the product at high prices and act as a broker for the cartel. The presence or absence of collective action depends on the consumers' sympathy for each other.

In this context, the uranium market is fairly well suited to cartelization, since a lot of countries plan to import uranium concentrates. However, the six leading importers—West Germany, Japan, Spain, the United Kingdom, France, and Sweden—represent a large share of total imports.

8. *Similar supply elasticities among producers.* When a cartel restricts production, prices rise and marginal costs fall. If one producer's supply curve is much less elastic than all the other producers' supply curves, he will receive a greater profit from the last (or highest-cost) unit of production, if production quotas are based on an equal percentage cutback from competitive production levels. The producer with inelastic supply will receive more than his fair share of the increase in all the producers' profits attributable to the cartel. In other words, whoever achieves the greatest cost reductions as a result of the cartel will achieve the greatest increase in profit per unit of output. Since the other members of the cartel will regard this as an inequitable distribution of cartel profits, they may try to establish a system of production quotas under which these profits are distributed in proportion to the market shares which would prevail in a competitive market. This system would be much more difficult to implement than equal percentage reductions in output. Obviously, the producers with excess profits would try to hide them from the other cartel members, and some system of independent auditing would have to be created. Standard accounting systems would be proposed; each producer

would favor a system which underestimated his true share of cartel profits. In short, differences in cost curves could lead to endless haggling and break up the cartel.

For a potential uranium cartel, this could be a major problem. Different producers mine a wide variety of deposits, at various depths, thicknesses, and ore grades, so it is reasonable to suspect that a wide range of supply elasticities exists. Moreover, costs are not simply a function of production rates; it is possible for most uranium mining companies to choose between mining the low-cost deposits soon and maximizing current profits or saving the low-cost deposits for later. A company can hide excess profits by leaving its low-cost reserves in the ground. It is difficult to imagine how any uranium cartel could set production quotas based on the concept of fair shares of the increase in profits attributable to the cartel. However, producers who believed that they were receiving less than their fair share might nevertheless decide not to complain about it, lest the cartel fall apart.

We use the word *profit* because it can be used to accurately describe a cartel among corporations. However, a cartel among nations might receive cartel profits in the form of taxes and royalties. The term *take* is sometimes used to describe revenues received by OPEC governments.

In summary, the present condition of the uranium market would be very favorable for cartel formation, but for the presence of two major obstacles: the ability of several new producers to enter the market at cartel prices and the impossibility of establishing an equitable distribution of cartel profits. To these difficulties we may add the general political differences between four groups of exporters: Australia and Canada, South Africa, the rest of Africa, and South America. The success of a uranium cartel during 1972–1974 can be attributed in large part to the fact that the latter two groups represented only a small share of the market, the French controlled production from Niger and Gabon, and the issue of nuclear proliferation had not divided Australia, Canada, France, and South Africa. On the whole, we believe that a successful uranium cartel would be difficult to establish in the future.

If a cartel were to form in the non-Communist, non-U.S. sector, it would face competition from U.S. producers unless one of two conditions prevailed: either the United States would rely on net imports, and domestic demand would exceed domestic supply at the cartel price level, or else the United States would impose an embargo on exports of domestically produced uranium concentrate. These conditions were discussed in chapter 7. In our view, one or the other is likely to prevail; that is, the United States is not likely to become a leading net exporter of yellowcake. In this sense, a cartel would not be threatened by the United States.

Table 8–7
Comparison of Free World Uranium Production with U.S. Purchases of Uranium, 1956–1976
(in thousand short tons U_3O_8 in concentrate)

Year	Reported Free World Production[a]	AEC Purchases Domestic[b]	AEC Purchases Foreign[a]	Deliveries to Commercial U.S. Buyers Domestic[c]	Deliveries to Commercial U.S. Buyers Foreign[c]	Remainder[d]
1956	14.470	5.958	7.500	—	—	1.012
1957	23.270	8.482	11.826	—	—	2.962
1958	36.250	12.437	16.500	—	—	7.313
1959	43.350	16.239	18.570	—	—	8.541
1960	41.130	17.637	15.770	—	—	7.723
1961	36.300	17.348	12.915	—	—	6.037
1962	34.500	17.008	11.720	—	—	5.772
1963	31.025	14.217	8.802	—	—	8.006
1964	26.204	11.846	5.297	—	—	9.061
1965	20.586	10.442	2.650	—	—	7.494
1966	19.520	9.488	2.048	0.2[e]	—	7.784
1967	19.098	8.425	—	0.7[e]	—	9.973
1968	23.005	7.337	—	4.85	—	10.868
1969	23.083	6.184	—	4.2	—	12.699
1970	24.161	2.520	—	9.3	—	12.341
1971	23.903	—	—	12.7	—	11.203
1972	25.647	—	—	11.6	—	14.047
1973	25.635	—	—	12.1	—	13.535
1974	24.176	—	—	11.9	—	12.276
1975	25.034	—	—	12.5	1.1	11.434
1976	30.650	—	—	13.8	2.9	13.950

[a]U.S. Department of the Interior, Bureau of Mines, *Minerals Yearbook* volumes for 1956 through 1974; and *Commodity Data Summaries 1977*. See table 8–2.

[b]USERDA, *Statistical Data of the Uranium Industry*, GJO-100(76) (Grand Junction, Colo.: USERDA, 1976), p. 12.

[c]Annual AEC and ERDA surveys of U.S. uranium marketing activity, 1968–1977.

[d]This column is obtained simply by subtracting AEC purchases and deliveries to commercial U.S. buyers from reported free world production.

[e]The AEC reported total deliveries of 900 tons in 1966–1967; the Bureau of Mines reported deliveries of 100 tons in 1966 and 700 tons in 1967.

One reason why it is difficult to predict future market behavior is that the uranium market has been commercial and competitive only in the last 10 years. Prior to 1968, a large share of production went to the U.S. Atomic Energy Commission, as table 8–7 shows. It was reported in 1969 that "since the start of the procurement program, the United States has purchased roughly 85% of the uranium produced in the western world, all of it until recently going to the AEC. Britain has bought most of the balance."[37]

The ability of developed nations to form a consumers' cartel in retaliation against a producers' cartel is a complicated topic which we shall not review here.

The Current Outlook for Uranium Prices

It is very difficult to obtain data on worldwide uranium price trends. The Uranium Institute does not report on uranium prices, lest it be accused of restraining competition. The chairman of the Institute observed in July 1977 that the topic is a very sensitive one. "We need a mechanism which avoids great oscillations in the price of uranium. . . . But at the moment, for one half of our industry, even to discuss the matter is to risk investigation."[38] An attempt to establish a price-reporting system was made at a meeting in late 1976:

> The meeting was the annual conference of the World Nuclear Fuel Market (WNFM), a "club" composed predominantly of uranium consumers, set up in 1974 by Nuclear Assurance Corporation with the declared objective of "being the world's marketplace for nuclear energy commodities." . . . As currently envisaged, monthly compilations of the data on all transactions would be sent to participants, and running averages of U.S. and non-U.S. prices would also be shown.[39]

This system would face not only antitrust problems, but a lack of participation by uranium producers and vertically integrated utilities.

It is possible to get a rough idea of the utilities' expectations about prices for uranium produced outside the United States by examining recent trends in import and export commitments. Although some large import contracts were signed in 1974, no such commitments were made in 1976. Moreover only 5,500 short tons U_3O_8—a small percentage of future domestic production—was committed for export as of January 1, 1977.[40] Consequently, we believe that the outlook for uranium prices in the world market is similar to the outlook for the domestic market, which we described in chapter 7: a modest annual rate of escalation, perhaps 2 percent above the increase in the U.S. consumer price index from a base of about $40 per pound U_3O_8 in 1977. The uncertainty surrounding any projection of world prices, however, is even greater than the uncertainty involved in any domestic forecast.

Notes

1. Atomic Industrial Forum, "Foreign Nuclear Electric Capacity Soars 33% in One Year, Now Surpasses U.S. Total," news release, June

10, 1977; A.M. Angelini and G.F. Castelli, "Integrated Planning of the Fuel Cycle," *Nuclear News*, July 1977, p. 56.

2. For example, see the April issue of *Nuclear Engineering International* for any year. Companies selling nuclear equipment and services in the United States are listed in the annual *Nuclear News Buyer's Guide*.

3. OECD Nuclear Agency and International Atomic Energy Agency, *Uranium: Resources, Production, and Demand* (Paris: OECD, 1977), p. 32.

4. Owe Carlsson, "The Ranstad Project and Other Swedish Projects and Possibilities," in *Uranium Supply and Demand*, edited by M.J. Spriggs and K.D. Casteel (London: Mining Journal Books, 1977), pp. 203–211.

5. The Bureau of Mines includes South West African production in the figures for South Africa.

6. Bernard C. Duval, "The Changing Picture of Uranium Exploration," in *Uranium Supply and Demand*, edited by M.J. Spriggs and K.D. Casteel (London: Mining Journal Books, 1977), pp. 126–127.

7. Field Movements of IAEA Experts," *International Atomic Energy Agency Bulletin* 19,4 (1977):58–59.

8. Norman Gall, "Atoms for Brazil, Dangers for All," *Foreign Policy* 23 (1976):160. See also Edward Wonder, "Nuclear Commerce and Nuclear Proliferation: Germany and Brazil, 1975," *Orbis* 21,2 (1977):296–297.

9. Duval, "The Changing Picture of Uranium Exploration," pp. 126–127.

10. "Field Movements of IAEA Experts," pp. 57–58; A.J. Polliart and P.M.C. Barretto, "Nuclear Energy Prospects and Uranium Resources in Latin America," *International Atomic Energy Agency Bulletin* 18, 3/4 (1976):23.

11. OECD Nuclear Energy Agency and IAEA, *Uranium: Resources, Production, and Demand*, p. 51.

12. Walter C. Woodmansee, "Uranium," in *Minerals Yearbook, 1971*, prepared by U.S. Department of the Interior, Bureau of Mines (Washington: GPO, 1973), p. 1219; also *Minerals Yearbook, 1972*, p. 1281.

13. "Europe Turns a Hungry Eye on Africa's Uranium," *African Development*, 9,8 (August 1975):17.

14. Gall, "Atoms for Brazil, Dangers for All," pp. 169–170.

15. For a discussion of fuel assurance, see Steven J. Warnecke, "Fuel assurance and supply security," *Uranium Supply and Demand*, proceedings of the Uranium Institute (London: Mining Journal Books, 1978), pp. 313–324.

16. U.S. General Accounting Office, *Certain Actions That Can Be*

Taken To Help Improve This Nation's Uranium Picture, EMD-76-1 (Washington: USGAO, July 2, 1976), pp. 11–13; George F. Combs, Jr. and John A. Patterson, "Uranium Market Activities," *Uranium Industry Seminar*, GJO-108(78) (Grand Junction, Colo.: USDOE, 1979).

17. Luxembourg can sell its surplus nuclear power to surrounding countries, but Iceland cannot.

18. See "Pakistani Nuclear Expert Asks $100 Billion Fund for Renewable Energy," *Energy Daily*, December 7, 1977, p. 6.

19. Bernard Fisk, "Australian Uranium Today," *Uranium Supply and Demand*, proceedings of the Uranium Institute (London: Mining Journal Books, 1978), pp. 95–103.

20. "Field Movements of IAEA Experts," p. 59; George Quester, *The Politics of Nuclear Proliferation* (Baltimore: Johns Hopkins, 1973), p. 216.

21. Owe Carlsson, "The Ranstad Project and Other Swedish Projects and Possibilities," in *Uranium Supply and Demand*, edited by M.J. Spriggs and K.D. Casteel (London: Mining Journal Books, 1977), pp. 203–211. By comparison, the amount of uranium ore received by mills in the United States in 1976 was 8.16 million metric tons: U.S. Energy Research and Development Administration, *Statistical Data of the Uranium Industry*, GJO-100(77), (Grand Junction, Colo.: USERDA, 1977), p. 100.

22. Quester, *The Politics of Nuclear Proliferation*, p. 152.

23. For a list of reprocessing plants, see Stockholm International Peace Research Institute, *World Armaments and Disarmament: SIPRI Yearbook 1977* (Cambridge, Mass.: M.I.T. Press, 1977), pp. 47–49.

24. R.E. Worroll, "The Pattern of Uranium Production in South Africa," in *Uranium Supply and Demand*, edited by M.J. Spriggs (London: Uranium Institute, 1976), p. 26.

25. Atomic Industrial Forum, "Foreign Nuclear Electric Capacity Soars 33% in One Year, Now Surpasses U.S. Total," news release, June 10, 1977.

26. "Europe Turns a Hungry Eye on Africa's Uranium," p. 18.

27. "Overplayed Hand?," *Forbes*, September 15, 1976, p. 54.

28. OECD Nuclear Energy Agency and International Atomic Energy Agency, *Uranium: Resources, Production, and Demand* (Paris: OECD, 1977), pp. 20–21.

29. Paul Lewis, "Schlesinger Warns West U.S. Energy Plan Is Vital," *New York Times*, October 6, 1977, p. 79.

30. Klaus-Peter Messer, "Uranium Demand as Judged by Electric Utilities," in *Uranium Supply and Demand*, edited by M.J. Spriggs and K.D. Casteel (London: Mining Journal Books, 1977), p. 61.

31. Milton R. Benjamin, "U.S. Officials View Pakistan as the Leading Threat to Join the Nuclear Club," *Washington Post*, December 8, 1978.

32. Carlsson, "The Ranstad Project," pp. 209–210.

33. U.S. Energy Research and Development Administration, *Statistical Data of the Uranium Industry*, GJO-100(77), p. 22.

34. See note 1.

35. The Central Intelligence Agency has published a report on Soviet oil production, and it is the agency most likely to issue a study of Soviet uranium production. See U.S. Central Intelligence Agency, *Prospects for Soviet Oil Production* (Washington: USCIA, April 1977).

36. This approach is used by Mason Willrich and Philip M. Marston in "Prospects for a Uranium Cartel," *Orbis* 19,1 (Spring 1975):166–184. However, we begin with a different set of theoretical conditions and examine a somewhat different set of facts.

37. "Report on Uranium," *World Business* 16 (July 1969):10–11.

38. Heinrich Mandel, "Concluding Remarks," in *Uranium Supply and Demand* (1977), pp. 247–248.

39. Simon Rippon, "Ways of Talking about Uranium Prices," *Nuclear News* (December 1976):42–43.

40. U.S. Energy Research and Development Administration, *Survey of United States Uranium Marketing Activity*, ERDA 77-46 (Washington: USERDA, May 1977), pp. 12–13; see also table 2–3.

9 Development and Use of the Analytical Model

In this chapter we will address three subjects: the development of a computer simulation model for the study of uranium leasing, the use of the model to describe hypothetical production centers, and the identification of data requirements needed to evaluate different leasing schedules.

Development of the Model

Our computer analysis of uranium leasing is based on modifications of a generalized resource policy evaluation model, called *GEN2*, which was developed at Cornell University. The model has been used to study the leasing of coal-bearing federal lands.[1] Earlier, less-sophisticated versions of the model were used to study the leasing of oil shale and outer continental shelf oil.[2] A modified version of GEN2 has been developed to study enhanced oil recovery on the Outer Continental Shelf.[3] Research on the leasing of geothermal resources is in progress.[4]

The GEN2 computer program contains several subroutines. The revisions necessary for adapting the model to uranium leasing were fairly simple. First, a subroutine to calculate uranium costs was added. Second, the subroutine which reads input data was modified so that a uranium data card would be read once for each control card specifying uranium as the resource to be analyzed. Finally, the rest of the program was revised so that the flow of logic would pass to and from the uranium cost subroutine.[5] A description of the model was published prior to these revisions.[6] In the following discussion, primary attention will be given to the aspects of the present model which are specifically related to uranium resources. However, we may begin by describing the general nature of the model.

General Characteristics

The simulation program calculates the values necessary to describe the cash flow of a single production center involved in resource extraction—for example, an offshore oil platform, a coal mine, a uranium mine-mill complex, or an oil shale facility. Discounted cash flows as well as annual values are calculated. By aggregating the cash flow of many production centers, one can estimate the cash flow for a major portion of the production centers to be developed in the United States in the future. For example, one might aggregate all the production centers expected to be built on lands leased by the federal government over the next 20 years—all the offshore oil platforms or all the coal mines, etc. However, the model is strictly a supply model and does not derive prices by simulating the interaction of supply and demand and federal price regulation. In studying uranium production, we have used the model only to describe individual mine-mill complexes, since the geological and ownership data needed for an aggregate picture are not available.

The GEN2 program can be used to describe the cash flow associated with a particular reserve deposit, a particular price projection, a particular operating-cost projection, and a particular investment-cost estimate. We shall call this set of calculations a *mean run*. One of the major strengths of the program is that it can describe the uncertainty faced by a private developer by calculating statistics which describe distributions of output values generated by a random sampling from a reserve distribution and/or a set of price, operating-cost, and investment-cost distributions. In other words, Monte Carlo simulation is used to model uncertainty. The reserve distribution can take on any of a variety of possible forms. The price distribution must be normal for any given point in time; the mean can change as a function of time. The two cost distributions must be triangular, that is, defined entirely by a minimum value, a most likely value (mode), and a maximum value for any given point in time; the mode can change as a function of time. In studying uranium, we have assumed that a given reserve deposit is known with certainty, and not subject to random sampling.

Separation of Exploration from Development

Our analysis of uranium mine-mill complexes is based on the development of a known reserve deposit. We have not attempted to model the leasing of unexplored uranium lands. There are four reasons for limiting the scope of our inquiry in this way.

1. Uranium exploration over a given tract of land is a much riskier activity than the development of a known deposit on a lease tract of similar size. A private company attempting to maximize its profits should regard the "product" of exploration to be the discovery of uranium deposits for which the after-tax net present value associated with development is positive. The exploration phase, from initial surveys to the discovery of such a deposit, should be evaluated by a discounted cash flow in which the level of risk is very high. The development phase should

be evaluated by another discounted cash flow. If risk is borne by the private firm, a higher discount rate should be used in the exploration phase, and the cash bonus offered for exploration rights should be a smaller fraction of the mean after-tax net present value than the cash bonus offered for development rights. In short, a two-phase modeling procedure is appropriate.

2. As we saw in chapter 3, the acreage of a uranium claim is very small—perhaps 100 acres, on average. The legal problems associated with consolidating tracts of land for uranium leasing are sufficiently great that large leaseholds, comparable to the 5,760-acre tracts defined by outer continental shelf oil leasing, are likely to be the exception. Consequently, the risk associated with exploration under a particular uranium lease is likely to be quite high. Since the federal government owns a vast amount of uranium-bearing land, it should absorb the exploration risk by receiving economic rent through contingency payments. For example, the government could lease land for exploration only and offer to pay the exploration company a percentage of the cash bonus (or the after-tax net present value of the bonus plus contingency payments plus federal taxes) which is received when a valuable deposit is leased for development. The ownership pattern of uranium lands dictates that the government will capture a greater portion of the economic rent if it issues separate leases for exploration and development in order to absorb risk in the exploration phase.

3. At present the government is not receiving the fair market value of uranium-bearing lands on which valuable deposits have already been discovered. Consequently, it is likely that a leasing system would extend to known deposits and not just unexplored territory. The most urgent policy questions are associated with these known deposits.

4. The data required to estimate the probability of finding a given type of deposit (characterized by size, grade, depth, and thickness) on a given tract of land are unavailable. Although the discovery of uranium reserves can be regarded as a process of random sampling from some sort of probability distribution, it is very hard to even begin to describe that distribution. For our purposes, three parameters are involved—the values of a, b, and c in equation 5.16. We would rather not attempt to define distributions for them.

Key Assumptions

The GEN2 simulation model has been developed to handle a wide variety of different leasing systems and to compute federal and state income taxes in some detail. Consequently, the program is long and complicated, and it is difficult to even draw a flowchart incorporating all the subroutines.

However, there are very few assumptions in the model which have the effect of making a statement about the decision-making behavior of the firm or of the landowner (in this case, the federal government). In our view, there are only five key assumptions in the version which we have used to study uranium resource policy:

1. The cutoff grade of ore is determined by equating the marginal cost per short ton of ore to the marginal after-tax revenue per short ton of ore, net of royalties, severance taxes, and federal and state income taxes. We stated this in equation 5.22. In effect, we have simply assumed that the firm maximizes after-tax profits.

2. On mean runs the firm is assumed to select the level of capacity which maximizes the after-tax net present value of the uranium reserve. The program calculates a cash bonus bid based on this maximum value. A choice among five different levels of capacity is assumed. It is possible to have the program print out annual and discounted cash flows for each level of capacity, so it is not too difficult to establish a different assumption about the choice of optimal capacity unless Monte Carlo iterations are involved. For Monte Carlo runs, the program uses the calculation of the cash bonus to determine whether or not development is to occur for a particular iteration. Development is assumed whenever the after-tax net present value of a possible project is greater than the tax writeoff associated with the cash bonus payment.

3. The program stores output statistics for each Monte Carlo iteration for the capacity level which maximizes after-tax net present value on that particular iteration. The program will print output statistics associated with the capacity level which is optimal for the largest percentage of the total number of iterations on a given run. In addition, statistics are printed for the capacity level which is optimal for the second largest percentage of the total number of iterations. It is theoretically possible, therefore, for the printed statistics to describe only 40 percent of the total number of iterations. We assume that these statistics will provide the information necessary to select the optimal bidding system, even though a complete set of Monte Carlo iterations is not computed for each of the five capacity levels.

4. Prices and costs are assumed to be exogenous. In other words, the demand for the resource is assumed to be perfectly elastic, and the supplies of labor, capital, and materials are also assumed to be perfectly elastic. These assumptions are reasonable when an individual production center is being studied, but not when a large number of production centers are aggregated. We have not used the model to study such aggregations.

5. Tax writeoffs are assumed to be used by the individual firm, regardless of the amount of these writeoffs. Values for federal and state income tax can be negative and are not constrained to exceed some level.

This assumption is equivalent to stating that the individual deposit is being developed by a firm which has other investments and is sufficiently profitable to owe federal and state taxes.

Calculation of ATNPV

We shall use the abbreviation ATNPV to refer to after-tax net present value. Since this is the most important output variable, we will briefly describe the way in which it is calculated. For each year of the lease up to and including the final year of production, the model will compute after-tax value as follows:

$$
\begin{aligned}
\text{After tax value} &= \text{gross revenue} - \text{operating cost} - \text{royalty} \\
&- \text{severance tax} - \text{rent} - \text{profit share} - \text{federal income tax} \\
&- \text{state income tax} = \text{taxable income} + \text{depletion} + \text{depreciation} \\
&- \text{federal income tax} - \text{state income tax}
\end{aligned} \tag{9.1}
$$

Then after-tax values are discounted to determine their present value, which is used to compute ATNPV. We assume a 4-year development period for uranium:

$$
\text{ATNPV} = \sum_{t=1}^{T+4} (\text{after tax value}) \int_{t-1}^{t} e^{-Rs} \, ds
$$

$$
- \sum_{t=1}^{4} (\text{annual investment cost}) \int_{t-1}^{t} e^{-Rs} \, ds \tag{9.2}
$$

$$
\begin{aligned}
&- \text{acquisition cost of lease, excluding the bonus} \\
&+ (\text{working capital} + \text{salvage value of initial investment}) \\
&\frac{1}{(1 + R)^{TT+4}} \\
&+ \text{present value of salvage of deferred investment}
\end{aligned}
$$

where R = discount rate
$T = TT$ = production lifetime

Open-pit mine development which occurs after the start of production is always included in "operating cost" in the equation for after-tax value. Open-pit or underground mine development which occurs before the start of production may be either capitalized or expensed. The input variable for "percent investment tangible" (a figure used in calculations for oil leasing) is used to describe the percent of preproduction uranium mine development which is capitalized.

Calculation of Gross Revenue and Production

There are a few other calculations which should be noted briefly. All these equations are based on the cost estimation procedure described in chapter 5:

$$F_{TT} = \frac{e(TT) \cdot R_c(TT)}{q \cdot 365 \text{ d/y}} \tag{9.3}$$

where F_{TT} = capacity factor for the final year of production
    $R_c(TT)$ = ore reserve which would exist in the original deposit if it were mined at the cutoff grade used in the final year of production, in ST ore
    $e(TT)$ = portion of $R_c(TT)$ which remains to be mined in the final year of production, expressed as a fraction per year
    q = capacity, in ST ore/d
    d   = day
    y   = year

$$
\begin{aligned}
y_o(tt) &= F \cdot q \cdot 365 \text{ d/y} \quad tt < TT \\
&= F_{TT} \cdot q \cdot 365 \text{ d/y} \quad tt = TT
\end{aligned}
\tag{9.4}
$$

where $y_o(tt)$ = annual ore production, in ST ore/y
    F = capacity factor in each year except the final year of production

$$y_u(tt) = y_o(tt) \cdot x(tt) \cdot \frac{G_a(tt)}{100\%} \cdot 2000 \text{ lb/ST} \tag{9.5}$$

where $y_u(tt)$ = annual uranium production for a given year, in lb U_3O_8/y
    $x(tt)$ = recovery rate for a given year
    $G_a(tt)$ = average ore grade for a given year, in percent U_3O_8 per unit of ore

$$L(tt) = y_o(tt) \cdot K \cdot e^{\theta t} \tag{9.6}$$

where $L(tt)$ = annual operating cost for a given year, in $/y
    K = initial operating cost, in $/ST ore
    θ = index of escalation for operating cost

$$I(tt) = y_u(tt) \cdot P \cdot e^{\Omega t} \tag{9.7}$$

where $I(tt)$ = annual gross revenue for a given year, in \$/y
$\quad P$ = initial price, in \$/lb U_3O_8
$\quad \Omega$ = index of escalation for uranium price

$$y_{CO} = [(TT - 1)F + F_{TT}]q \cdot 365 \text{ d} \tag{9.8}$$

where y_{CO} = cumulative ore production, in ST ore

$$y_{CUD}(tt) = \frac{y_u(tt)}{e(tt)}$$
$$= R_c(tt) \cdot x(tt) \cdot \frac{G_a(tt)}{100\%} \cdot 2000 \text{ lb/ST} \tag{9.9}$$

where $y_{CUD}(tt)$ = estimate of cumulative uranium production made for the purpose of calculating depreciation in a given year, in lb U_3O_8

For capital investments made prior to the start of production, excluding mine equipment, depreciation can be calculated in relation to the rate of resource depletion. This procedure allows a reasonable estimate of depreciation to be computed for a wide range of production lifetimes, even though the production profile is not predictable when prices and operating costs are subject to random variations. Of course, cumulative depreciation is constrained to equal the original investment. Mine equipment is replaced every 5 years and depreciated over 5 years.

Calculations of Mineral Inventory Values

We have selected values for the geological parameters a, b, and c so that they describe a hypothetical deposit which we consider to be typical of large sandstone deposits in the United States. Unfortunately, we lack empirical data on mineral inventories for individual deposits.

If equation 5.19 is solved for a and that expression is substituted for a in equation 5.20, the following relationship is found to exist:

$$\frac{G_a + c}{G_c + c}\left(1 - \frac{R_o}{R_c}\right)(b + 1) = 1 - \left(\frac{R_o}{R_c}\right)^{b+1} \tag{9.10}$$

If R_o is much smaller than R_c, the ratio of R_o to R_c is negligible, and the equation can be simplified:

$$(G_a + c)(b + 1) = G_c + c \tag{9.11}$$

This equation, in turn, can be solved for b:

$$b = \frac{G_c - G_a}{G_a + c} \qquad (9.12)$$

If we assume that $R_o = 1$ ST ore, then

$$G_o = a \cdot R_o^b - c = a - c \qquad (9.13)$$

In other words, the cutoff grade of the first ton of ore is equal to $a - c$. Let us assume that $a = 100$ percent, $G_c = 0.015$ percent, and $G_a = 0.08$ percent. Selection of different values of c will yield different values of b, given equation 9.12, and different values of R_c, given equation 5.19. If we assume $c = 0.06$, we find that $R_c = 5{,}378{,}204$ ST ore. This is a fairly large deposit, but not extraordinary; there were forty-three domestic deposits holding \$30 reserves of more than 5 million ST ore as of January 1, 1977.[7]

The uranium reserve associated with an ore deposit can be calculated simply:

$$U_c = (R_c - R_o) \frac{G_a}{100\%} \qquad (9.14)$$

where U_c = uranium contained in ore mined from a given deposit down to a cutoff grade equal to G_c, in ST U_3O_8

Note that equation 5.12 must be used to calculate the amount of uranium production, that is, the amount which is actually recovered in the form of yellowcake.

Use of the Model

We have used the GEN2 simulation program to study three different topics: the response of an individual production center to price changes, the distribution of after-tax net present value for a single capacity level under price and cost uncertainty, and the performance of different leasing systems. All the computations have been based on the common input values shown in table 9–1. As was noted in chapter 5, the open-pit production costs assume a depth-to-thickness ratio of 24 and the underground production costs assume a depth-to-thickness ratio of 76.

Table 9–1
Common Input Values for Leasing-Policy Analysis

Price related	
Mean of price change distribution	2%
Standard deviation of annual price change distribution	5%
Tax related	
Depreciation method for original investment	Rate of resource depletion
Depreciation life for deferred investment	5 years
Percent original investment salvageable	20%
Percent deferred investment salvageable	10%
Federal corporate income tax rate	48%
Investment tax credit rate	10%
State severance tax rate	4%
State corporate income tax rate	4%
Time related	
Minimum production time	1 year
Maximum production time	25 years
Length of development period	4 years
Mine-mill capacity factor (for all years except the final year of actual production)	0.80
Cost related	
Triangular investment and operating cost contingency distributions	
Minimum	−0.10
Most likely	0
Maximum	0.30
Rent per acre	0
Annual rate of change in operating cost	1%
Percent of development cost which is not expensed ("tangible")	0%
Working capital factor—multiplied by first year's operating cost to determine working capital value	0.10
Lease acquisition cost (cost of bid preparation)[a]	0
Exploration cost[a]	0
Geological	
a	100
b	−0.4643
c	0.06
Other factors	
Bonus factor—multiplied by after tax net present value to determine bonus	0.75
Number of Monte Carlo iterations	200

Note: Selection of the optimal level of production capacity was based on a choice among five different levels: 1,500, 2,000, 2,500, 3,000, and 3,500 ST U_3O_8 per day.

[a]These costs are excluded from our analysis, but they are not negligible.

Estimate of Supply Curves

To evaluate the supply response of an individual production center, we assumed a straight cash bonus bidding system with no royalty, a discount rate of 8 percent, a depletion allowance of 23 percent, and a tailings management cost of 25 cents per short ton of ore. The results are presented in figure 9–1; these are estimates of cumulative uranium concentrate production, assuming mean values of prices, operating costs, and investment cost. In figure 9–1 the dashed lines indicate the price ranges at

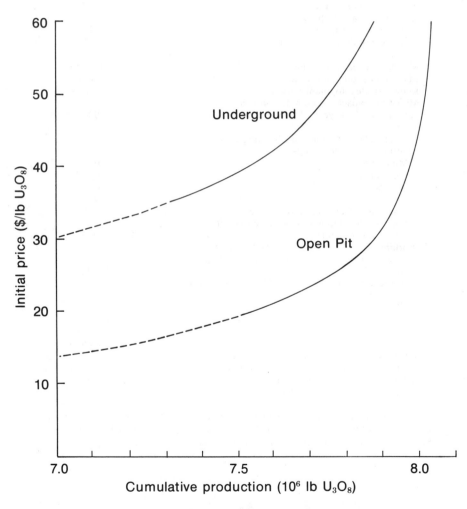

Figure 9–1. Supply Curves for Individual Mine-Mill Production Centers.

which ATNPV is negative. These results show that real price increases above the current level of around \$40/lb U_3O_8 will bring forth an increase in the quantity of uranium produced at a large underground mine, but will yield practically no increase in production from a large open-pit mine. The basic explanation for this pattern is quite simple: operating costs per short ton of ore are higher at underground mines. The data in figure 9–1 also show that open-pit mines yield more economic rent than underground mines of comparable size.

Monte Carlo Simulation for a Single Level of Capacity

Our primary objective in applying the GEN2 model to uranium supply is to evaluate alternative leasing systems by Monte Carlo simulation. As noted earlier, the model does not generate a complete set of Monte Carlo iterations for each of the capacity levels described in the input data. We will use 200 iterations—the maximum permitted by the storage capacity of the model—to describe choices among five levels of capacity. As a preliminary step, we have constrained the choice to a single level of capacity in order to know what kind of probability distributions for ATNPV generate the sample distributions we will use to evaluate leasing systems. This exercise has been performed for a single set of price and cost assumptions, for a straight cash bonus bid system with no royalty.

Implicit in this preliminary analysis is the assumption that a firm will select a capacity level and then decide—sometime during the first two years of the lease—whether or not development is justified. No development occurs in each Monte Carlo iteration in which the ATNPV associated with development is less than the tax writeoff associated with a cash bonus equal to 75 percent of the ATNPV calculated on the mean run. In such situations the ATNPV for the Monte Carlo iteration is assumed to equal the tax writeoff.

In effect, the decision whether or not to develop a lease is modeled as though the firm could anticipate its cash flow with perfect certainty. An objection may be raised that such a procedure is not only unrealistic but conflicts with one of the basic objectives of the program—to analyze uncertainty. Four points can be made in response to this objection:

1. The program deals with uncertainty by describing the array of possible outcomes facing the government at the time when a choice of leasing systems must be made and a decision whether to lease the deposit must be made.

2. The private firm may adopt one of three strategies regarding the decision to develop the lease. It may try to avoid the mistake of not developing a lease when development would have been justified. This

mistake can be avoided by accepting the risk of having an ATNPV less than the tax writeoff, and by establishing a high discount rate to compensate for this risk. Alternately, the firm may try to avoid the mistake of developing a lease when development does not turn out to be justified. This can be avoided by selecting only the "safe" investments for which the ATNPV is very likely to exceed the tax writeoff, and by establishing a relatively low discount rate. Last, the firm may assign equal weight to both types of mistakes and not try to avoid either one at the expense of making the other. This policy would be associated with an intermediate discount rate. We do not know which strategy a firm would pursue, nor do we know how successful it would be in meeting its objectives. In effect, the GEN2 model assumes the third strategy and a high degree of success. This is an arbitrary assumption, but it is probably no worse than any other.

3. The decision to develop a lease depends on the level of the cash bonus, which has been estimated somewhat arbitrarily. A risk-averse firm will offer a low cash bonus and thereby be more likely to expect an ATNPV above the tax writeoff. A risk-loving firm will offer a high cash bonus and thereby be more likely to expect an ATNPV *below* the tax writeoff. There is no sense in trying to model the development decision precisely when the cash bonus estimate is already so imprecise.

4. The government will select a leasing method and the firm will select a capacity level which maximize the likelihood that development is justified, *ceterus paribus*. In some instances, ATNPV is virtually certain to exceed the tax writeoff. In other instances, the ATNPV associated with development will fall below the tax writeoff in only a small percentage of the total number of Monte Carlo iterations, so that the mean and variance statistics will hardly be affected by the lack of consideration of the two types of mistakes just mentioned. As long as the GEN2 model can accurately identify the optimal leasing system and capacity level, we need not be concerned about a bias in favor of the suboptimal choices.

The results of our preliminary analysis are shown in figures 9–2 and 9–3. If development occurs on every Monte Carlo iteration, the distribution of ATNPV approximates a normal distribution with a mean equal to the ATNPV calculated on the mean run. The mean of the sample, given 200 iterations, is close to the ATNPV calculated on the mean run. If development does not always occur, the distribution of ATNPV approximates a truncated normal distribution, as in figure 9–3. When the sample includes only a small number of cases in which development occurs—say, less than 100—the distribution is very "noisy" and does not resemble any recognizable random distribution such as the normal distribution. Moreover, the mean of the sample will be significantly higher than the ATNPV calculated on the mean run. The sample mean is not very informative,

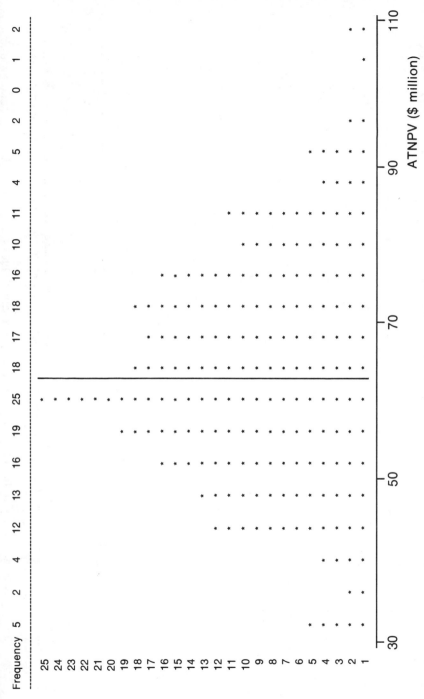

The calculations are based on simple cash-bonus bidding, a capacity of 2,500 ST ore/d, a price of $40/lb U_3O_8, a tailings disposal cost of 25 cents per short ton of ore, and a discount rate of 8 percent applied to real dollars. The vertical line indicates the after-tax net present value associated with the mean price, mean investment cost, and mean operating cost. Values are measured in millions of 1977 dollars.

Figure 9–2. Frequency Distribution for the After-Tax Net Present Value of a Lease for Open-Pit Production.

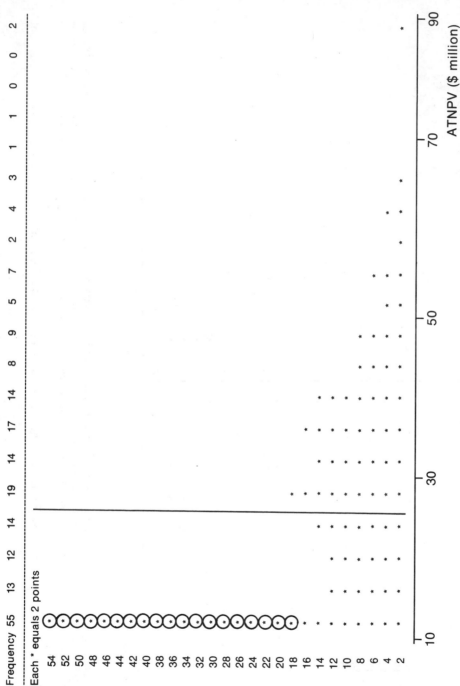

The calculations are based on the assumptions listed in figure 9–2. The circled asterisks indicate situations in which no development occurs. The vertical line indicates the after-tax net present value associated with the mean price, mean investment cost, and mean operating cost. Values are measured in millions of 1977 dollars.

Figure 9–3. Frequency Distribution for the After-Tax Net Present Value of a Lease for Underground Production.

however, since it does not suggest where the mean of the ATNPV distribution would be if it were not truncated.

These results show that if development occurs on every iteration for two different runs based on different input values, it is reasonable to assume that the sample values of ATNPV are taken from normal distributions. Consequently, it is simple to construct a statistical test to see whether the difference in the sample means is caused by a difference in the means of the underlying distributions. If we relax the assumption of a single capacity level and compute statistics for any of five capacity levels, this sort of test of statistical significance seems reasonable as long as each sample involves a capacity level which maximizes ATNPV over a large percentage of the total number of iterations. However, testing for statistically significant differences in sample means is not feasible when iterations frequently result in no development or when ATNPV is only occasionally maximized by any particular capacity level.

It can be seen that the sample distribution in figure 9–3 is flatter than the one in figure 9–2. This result suggests that the ratio of the mean of an ATNPV distribution to its standard deviation tends to be higher for open-pit mines than for underground mines, given identical input data. This hypothesis is confirmed by the data presented in table 9–2.

The calculations described in figures 9–2 and 9–3 were repeated for two other lease bidding systems: a cash bonus bid based on a 20 percent royalty and a cash bonus bid based on a 75 percent profit share. In the latter case, the federal government raises 75 percent of taxable income, as defined by the Internal Revenue Service, before income taxes are paid. In both cases the ATNPV distributions conformed to the general pattern we have described—normal or truncated normal, with a higher mean/standard deviation ratio for open-pit mines.

Performance of Different Leasing Systems

There are five basic types of payment by which economic rent can be transferred from a lessee to a lessor: a cash bonus, a royalty, a profit share, a production share, or an annual rental. In theory, any one of these could be the bid variable, but in practice, competitive bidding for leases on federal lands has involved the cash bonus as the bid variable (with the exception of a few offshore oil and gas leases involving royalty bidding). We have evaluated cash-bonus bidding under three alternatives: a "straight" cash bonus with no contingency payments, a fixed royalty of 20 percent, and a fixed profit share of 75 percent. At present there is little or no debate over alternative uranium lease bidding systems, so there is no need to review proposals which have been advocated by various organizations. Rather, it is necessary to evaluate the three alternative systems under different assumptions regarding the price of uranium concentrates, the cost of reducing environmental pollution from uranium mill tailings, the removal of the depletion allowance for uranium mining, and the choice of discount rate. We have limited the number of assumptions to two for each variable, so that the evaluation of open-pit and underground mines will involve altogether 96 runs of 200 Monte Carlo iterations, each involving five capacity levels.

The depletion allowance has the effect of expanding the quantity of uranium ore produced from a given deposit, as figure 9–4 shows. This figure is simply a graphical illustration of equation 5.22. In the upper half of figure 9–4, the variable z is assumed to be positive, and in the lower half, z is assumed to equal zero. Removal of the depletion allowance is comparable to imposition of a sort of constrained royalty. We have chosen to consider the presence or absence of the depletion allowance as a leasing policy variable rather than just an assumption about mining costs and accounting systems.

As noted in chapter 3, the overall policy objectives of the Interior Department with regard to existing mineral leasing programs are threefold:

1. To ensure an orderly and timely development of the resource in question
2. To protect the environment
3. To ensure the public a fair market value return on the disposition of its resources.

It is logical to use these objectives as the basis for establishing criteria for the comparison of leasing systems. In our view, the major impact of environmental protection policies on uranium leasing will be the way in which regulations regarding tailings management will be reflected in increased costs. The cost of tailings disposal can be treated as an input parameter, for which we can assign only a range of possible values.

The other two policy objectives can be furthered by evaluating leasing systems according to the following criteria:

1. The probability that leasing will result in development should be high.
2. The quantity of uranium concentrate produced should be large, whenever the lease is developed.
3. The ratio of the mean of the ATNPV distribution to its standard deviation should be high, in order to demonstrate that risk is being transferred to the public sector.

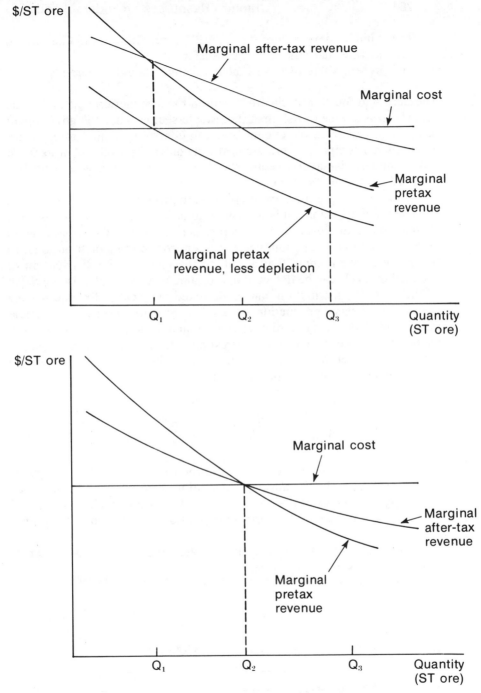

Figure 9–4. Effect of the Depletion Allowance on Production.

4. The standard deviation of the ATNPV distribution should be low, to demonstrate that risk is being transferred.
5. The present value of government revenue should be large.

The third and fourth criteria are based on the assumption that the public sector is less averse to risk than the private sector, since the government has so many sources of revenue and can control its deficit to a greater extent than a competitive private firm can control its profits. A transfer of risk to the less risk-averse sector should increase the total economic rent obtained from a particular leasehold.

In order to apply the second criterion to production figures generated by random distributions, it is necessary to construct a statistical test. We shall assume that if each of two samples is restricted to values associated with development, these samples are taken from normal distributions. In other words, we assume that a shift in the mean of the distribution of production levels can be detected in the same way that a shift in the mean of the ATNPV distribution can be detected. Typically, for one leasing system we observe production values X over n iterations, with sample mean $\bar{X}$ and sample standard deviation s_x, and for another leasing system we observe Y over m iterations, with simple mean $\bar{Y}$ and sample standard deviation s_y. According to probability theory, the difference between the sample means is normally distributed:[8]

$$\bar{X} - \bar{Y} \sim N\left(\mu_x - \mu_y, \ \sqrt{\frac{\sigma_x^2}{n} + \frac{\sigma_y^2}{m}}\right) \qquad (9.15)$$

where μ_x = mean of the normal distribution from which X is sampled
μ_y = mean of the normal distribution from which Y is sampled
σ_x^2 = variance of the normal distribution from which X is sampled
σ_y^2 = variance of the normal distribution from which Y is sampled

Therefore, the following equation holds true, because of the properties of the normal distribution:

$$P\left(\bar{X} - \bar{Y} > 1.28 \ \sqrt{\frac{\sigma_x^2}{n} + \frac{\sigma_y^2}{m}}\right) = 0.1 \qquad \text{when } \mu_x = \mu_y \quad (9.16)$$

We can estimate the true variances as follows:

$$\sigma_x = s_x \sqrt{\frac{n-1}{n}} \qquad \sigma_y = s_y \sqrt{\frac{m-1}{m}} \qquad (9.17)$$

We would like to test the following hypothesis, given that the X values are associated with a straight cash bonus bidding system:

$$H_o: \mu_x - \mu_y = 0 \qquad \text{versus} \qquad H_1: \mu_x - \mu_y > 0 \qquad (9.18)$$

The statistical test is quite simple; reject H_o if

$$\bar{X} - \bar{Y} > 1.28 \sqrt{\frac{s_x^2}{n-1} + \frac{s_y^2}{m-1}} \qquad (9.19)$$

Equation 9.16 tells us that when the true standard deviations are estimated correctly, the use of this test will result in only a 10 percent chance that we will believe the true means to be different when in fact they are equal.

The results of six Monte Carlo runs are shown in table 9–2. The presence or absence of the depletion allowance is treated as a leasing policy variable. The line marked Percent Developed shows the percentage of Monte Carlo iterations for which development occurs under each leasing system. The Change and Percent Change calculations compare the contingency systems to the straight cash bonus—with or without depletion, as appropriate. Changes in the sample means of the ATNPV distributions were always found to be statistically significant. The Most Likely capacity is the one which maximizes ATNPV over the largest number of iterations. Total Economic Rent consists of ATNPV plus the present value of taxes, royalty, and profit share; we estimate the mean of this sum by calculating the sum of the means for the four output variables. Government Revenue consists of the cash bonus plus the present value of taxes, royalty, and profit share; again, we estimate the mean of this sum. All the mean values refer to the iterations for which the Most Likely capacity maximizes ATNPV.

Statistics related to the first, third, and fifth evaluation criteria are summarized in tables 9–3 and 9–4. Data on production are omitted from these tables because there was found to be no significant tradeoff between this criterion and the three which are included. Production is maximized whenever the preferred system (marked by an asterisk) is a cash bonus or profit-share system with depletion. In two instances the profit share without depletion is preferred, and production is only slightly lower than it would be under a cash bonus with depletion. In three instances the cash bonus without depletion is preferred, and production is 1.5 to 1.6 percent below the level associated with depletion.

Data on the standard deviations of the ATNPV distributions are omitted from tables 9–3 and 9–4 because there are only two instances in

Table 9–2
Results of Monte Carlo Simulation of Alternate Leasing Systems[a]

	Cash Bonus		20% Royalty		75% Profit Share	
	with depl.	without depl.	with depl.	without depl.	with depl.	without depl.
Percent developed:	100	100	100	100	100	98.0
Change:			0	0	0	−2.0
Percent for which the most likely capacity maximizes ATNPV:	80.0	79.0	63.0	56.0	73.0	77.0
Installed capacity (most likely):	3500	3500	3500	3500	3500	2000
Installed capacity (mean run):	3500	3500	3500	3000	3500	3500
Years in production (mean run):	7	6	6	6	7	6
Cash bonus (based on mean run):	48.92	32.56	29.83	18.76	32.37	2.83
PV of federal tax (mean run):	19.68	44.00	10.05	26.67	−0.69	5.39
PV of state tax (mean run):	10.28	12.42	7.79	9.07	8.51	9.07
ATNPV (mean):	70.28	59.02	48.06	34.76	45.46	6.10
Percent change:			−31.62	−41.10	−35.32	−89.66
Mean/standard dev. ratio:	4.70	5.19	3.89	3.35	5.52	2.47
PV of taxes (mean):	36.92	53.06	22.29	34.39	10.14	15.67
PV of royalty (mean):			44.06	43.28		
PV of profit share (mean):					53.49	75.63
Total economic rent (mean):	107.20	112.08	114.41	112.43	109.09	97.40
Percent change:			+6.73	+.31	+1.76	−9.14
Government revenue (mean):	85.84	85.62	96.18	96.43	96.00	94.13
Percent change:			+12.05	+12.63	+11.84	+9.94
Production if developed (mean):	7.97	7.91	7.91	7.79	7.97	7.89
Percent change:			−0.75	−1.52	0	−0.25
Statistically sign. change:			Yes	Yes	No	Yes

Note: Fifteen other tables similar to this one were prepared, for different assumptions about mining method, initial price, tailings cost, and discount rate. All these tables are summarized in table 9–3.
[a]Mining method: open pit; initial price: $40 per lb U_3O_8; tailings cost: $.25 per ST ore; discount rate: 8%.

which this criterion plays an important role. When the profit-share system without depletion is preferred, the mean/standard deviation ratio for ATNPV is not maximized, but the level of risk faced by the private firm is not very great because the standard deviation is low.

At an initial price of $40 per pound U_3O_8 and a tailings cost of $6.00 per short ton of ore, none of the leasing systems yields a cash bonus for the underground deposit, regardless of the discount rate. Such a deposit could not be leased by competitive bidding, but it is conceivable that a lease would be issued in the interest of maximizing development of domestic uranium resources. Under these conditions, a straight cash-bonus system with depletion is preferred. The policy conclusion which we derive from these data is that the imposition of strict environmental

Table 9–3

Summary of Results of Monte Carlo Simulation for Open-Pit Mines.

Initial price Tailings cost Discount rate	Cash Bonus		20% Royalty		75% Profit Share	
	with depl.	without depl.	with depl.	without depl.	with depl.	without depl.
$40/lb U₃O₈						
$.25/ST ore						
8 percent						
Percent developed:	100	100	100	100	100*	98.0
Mean/st. dev. ratio:	4.70	5.19	3.89	3.35	5.52	2.47
Percent change in government rev.:			+12.05	+12.63	+11.84	+9.94
12 percent						
Percent developed:	100	100	99.5	99.5	100*	70.5
Mean/st. dev. ratio:	4.69	4.70	3.62	2.94	5.21	0.85
Percent change in government rev.:			+14.03	+13.26	+11.53	−25.38
$6.00/ST ore						
8 percent						
Percent developed:	99.5	100*	96.0	97.0	99.5	85.5
Mean/st. dev. ratio:	3.05	4.34	2.61	1.68	3.00	1.33
Percent change in government rev.:			+11.62	+3.94	+10.53	+2.29
12%						
Percent developed:	99.5	99.5*	93.5	92.5	99.5	40.0
Mean/st. dev. ratio:	3.06	3.75	2.27	1.34	2.93	0.50
Percent change in government rev.:			+12.05	−1.29	+11.56	−51.61
$60/lb. U₃O₈						
$.25/ST ore						
8 percent						
Percent developed:	100	100	100	100	100	100*
Mean/st. dev. ratio:	5.87	6.35	5.09	5.60	6.29	4.73
Percent change in government rev.:			+6.45	+9.70	+8.97	+10.56
12 percent						
Percent developed	100	100	100	100	100	100*
Mean/st. dev. ratio:	5.76	5.99	5.05	5.38	6.16	4.16
Percent change in government rev.:			+7.25	+9.77	+8.91	+34.73
$6.00/ST ore						
8 percent						
Percent developed:	100	100	100	100	100*	100
Mean/st. dev. ratio:	5.21	5.84	4.59	4.34	6.05	4.06
Percent change in government rev.:			+9.18	+9.62	+13.85	+6.15
12 percent						
Percent developed:	100	100	100	100	100*	99.5
Mean/st. dev. ratio:	5.35	5.43	4.48	4.30	6.06	3.11
Percent change in government rev.:			+9.88	+10.50	+9.84	+5.37

*This is the preferred leasing system.

requirements for uranium mill tailings will cause uranium prices to increase, because the cost of compliance will exceed the amount of economic rent available from most underground deposits at current price levels. Without a price increase, uranium from these deposits would not be supplied to utilities.

Table 9–4
Summary of Results of Monte Carlo Simulation for Underground Mines

Initial price Tailings cost Discount rate	Cash Bonus		20% Royalty		75% Profit Share	
	with depl.	*without depl.*	*with depl.*	*without depl.*	*with depl.*	*without depl.*
$40/lb U₃O₈						
$.25/ST ore						
8 percent						
Percent developed:	92.0	85.5	52.5	24.5	89.0*	1.5
Mean/st. dev. ratio:	1.57	1.28	0.58	0.32	1.45	0.14
Percent change in government rev.:			−34.89	−62.79	+14.50	−93.96
12 percent						
Percent developed:	88.0*	72.5	38.0	10.5	81.5	0
Mean/st. dev. ratio:	1.46	0.89	0.45	0.19	1.08	
Percent change in government rev.:			−34.75	−76.29	−10.23	
$6.00/ST ore						
8 percent						
Percent developed:	70.5*		9.5		59.0	
Mean/st. dev. ratio:	0.77		0.16		0.71	
Percent change in government rev.:			−54.16		+92.98	
12 percent						
Percent developed:	57.0*	28.5	6.0	1.5	43.0	0
Mean/st. dev. ratio:	0.61	0.34	0.12	0.11	0.53	
Percent change in government rev.:			−60.95	−86.80	+112.41	
$60/lb U₃O₈						
$.25/ST ore						
8 percent						
Percent developed:	100	100	99.5	99.5	100*	78.5
Mean/st. dev. ratio:	4.26	4.34	2.99	2.61	4.67	1.09
Percent change in government rev.:			+1.26	+.09	+13.62	−1.30
12 percent						
Percent developed	100	100	98.0	96.0	100*	17.0
Mean/st. dev. ratio:	4.16	3.81	2.88	1.91	4.74	0.30
Percent change in government rev.:			+6.45	−18.00	+19.39	−79.04
$6.00/ST ore						
8 percent						
Percent developed:	99.5	99.5	93.0	90.5	99.5*	52.5
Mean/st. dev. ratio:	3.52	3.65	1.70	1.47	3.63	0.60
Percent change in government rev.:			−13.82	−17.72	+10.88	−39.62
12 percent						
Percent developed:	99.5	99.5*	90.5	80.0	98.5	6.5
Mean/st. dev. ratio:	3.51	3.67	1.59	1.08	3.58	0.15
Percent change in government rev.:			−6.59	−37.74	+13.41	−91.88

*This is the preferred leasing system.

Our findings are summarized in table 9–5. No single leasing system is appropriate for all uranium deposits under all circumstances. Fixed-royalty systems are never preferred. It appears that the remaining alternatives can be ranked for either open-pit or underground production at a given discount rate according to the total economic rent associated with the deposit. The ranking is as follows:

Table 9–5
Summary of Recommendations Regarding the Choice of Leasing System[a]

	Open-Pit Production		Underground Production	
	Low Discount Rate	*High Discount Rate*	*Low Discount Rate*	*High Discount Rate*
High price, low tailings cost	Profit share, without depletion			
	188.92	173.55	168.05	123.92
High price, high tailings cost	186.02	137.19	124.67	Cash bonus, without depletion
		Profit share, with depletion		106.68
Low price, low tailings cost	109.09	76.95	39.87	31.14
Low price, high tailings cost	Cash bonus, without depletion		Cash bonus, with depletion	
	90.06	66.05	16.20	7.35

[a]Each figure represents an estimate of the mean value of the total economic rent at the capacity level which maximizes ATNPV most frequently under the particular conditions shown in the table.

1. A profit-share system without the depletion allowance is preferred for the very high-rent deposits.
2. A profit-share system with the depletion allowance is preferred for the moderately high-rent deposits.
3. A straight cash-bonus system without the depletion allowance is preferred for the moderately low-rent deposits.
4. A straight cash-bonus system with the depletion allowance is preferred for the very low-rent deposits.

Data Requirements for the Evaluation of Leasing Schedules

As was observed in chapter 3, the present legal system for disposing of uranium resources on federal lands is complicated and antiquated. Competitive bidding on mining claims does not exist, and it is possible for a private company or individual to withhold uranium from the market at a very modest out-of-pocket cost. It might be argued, therefore, that if a uranium-leasing system were introduced, the institutional obstacles to development might be reduced so substantially that uranium lands would be "dumped" on the market and the government would not receive fair market value for them. In this situation it would be necessary for the government to select an optimal leasing schedule.

The data requirements for forecasting the effects of different leasing schedules are formidable. Maps of surface and subsurface ownership would be required, together with maps of present uranium reserves and potential resources. The scale of such maps should permit descriptions of 160-acre sections, if not smaller areas. The relevant scale would depend on the ability of the government to consolidate tracts of land prior to leasing. Cooperation between the Bureau of Land Management, which receives data on mining claims, and the Grand Junction Office of the Department of Energy, which receives data on uranium reserves and drilling activity, is necessary to compile a reliable record of the extent and location of leasable uranium lands.[9]

In our view, it is unlikely that the government will develop a uranium-leasing policy under which lands are withheld from leasing in order to "save" them for future lease awards. The objective of maximizing uranium production is likely to be paramount in the selection of leasing procedures and schedules. However, environmental protection may have a major impact on lease schedules via Nuclear Regulatory Commission deliberations, which are subject to close public scrutiny.

Notes

1. Wallace E. Tyner, Robert J. Kalter, and John P. Wold, *Western Coal: Promise or Problem?*, Department of Agricultural Economics Research Paper 77-13 (Ithaca, N.Y.: Cornell Univ. Press, 1977).

2. Robert J. Kalter and Wallace E. Tyner, "An Analysis of Federal Leasing Policy for Oil Shale Lands," paper prepared for the Office of Energy R & D Policy, National Science Foundation, April 1975; Robert J. Kalter, Wallace E. Tyner, and Daniel W. Hughes, *Alternative Energy Leasing Strategies and Schedules for the Outer Continental Shelf*, Department of Agricultural Economics Research Paper 75-33 (Ithaca, N.Y.: Cornell Univ. Press, 1975).

3. Jeffrey Bodington completed a master's thesis on this topic at Cornell University in 1978.

4. John Broderick has worked on this subject.

5. The program is written in Fortran. All the programming associated with uranium was done by John Broderick.

6. Wallace E. Tyner and Robert J. Kalter, *A Simulation Model for Resource Policy Evaluation*, Department of Agricultural Economics Staff Paper 77-28 (Ithaca, N.Y.: Cornell Univ. Press, 1977). This paper was reprinted as chapter 3 of Tyner, Kalter, and Wold, *Western Coal: Promise or Problem?*

7. U.S. Energy Research and Development Administration, *Statistical Data of the Uranium Industry*, GJO-100(77) (Grand Junction, Colo.: USERDA, 1977), p. 59.

8. This discussion is based on Paul Meyer, *Introductory Probability and Statistical Applications* (Reading, Mass.: Addison-Wesley, 1972), pp. 325–326. See also Kalter, Tyner, and Hughes, *Alternative Energy Leasing Strategies and Schedules for the Outer Continental Shelf,* pp. 105, 108.

9. Data on the distribution of uranium resources by land status are found in U.S. Department of Energy, *Statistical Data of the Uranium Industry,* GJO-100 (78) (Grand Junction, Colo.: USDOE, 1978), p. B-3.

10 Conclusions

The preceding chapters have been devoted primarily to legal and technical questions, for which one can pull together facts and come up with answers or with pleas for more facts. In this chapter we would like to step back from the process of developing answers, and comment on the questions. There certainly is a lot of evidence that the nuclear industry has run into trouble by failing to ask the right questions. It has failed to anticipate the obstacles and criticisms which have arisen. In our view, this tendency is as strong in the context of uranium supply as in any other. We would like to venture a few conclusions, in the light of this observation.

First of all, we believe that nuclear fission is unlikely to become a leading source of energy in the United States. Nuclear power will form a contribution to our energy supplies, but there probably will not be a breeder-reactor economy—a civilization subsisting on plutonium. It follows that nuclear energy will be some sort of bridge to the future, something to tide us over while oil and gas become expensive.

If this is true, the uranium mining and milling industry will have a distinctly different role from the reactor manufacturing industry. The miners will have to keep reactors running when the nuclear industry is declining and being phased out of existence. This task could become very difficult if the tide of public opinion turns against nuclear power.

The key question regarding the supply of fuel for utilities may not be the geological availability of uranium, but the political and economic availability of uranium-bearing land. At present, the Mining Law of 1872 permits the industry to take land for its own use without paying for it. We believe that the industry may be making a serious error in judgment by trying to preserve this system.

Finally, we would like to point out that conflicts may arise between domestic and foreign policies regarding uranium. Neither the Bureau of Land Management of the Department of the Interior nor the Forest Service of the Department of Agriculture will be eager to increase the supply of uranium in order to discourage the use of plutonium abroad. The United States may be placed in the awkward position of withholding uranium resources while telling other countries not to use plutonium.

We have tried to put together a study of uranium supply which will age slowly. We will be pleased if it does, but if it does not, well, there will probably be a lesson worth learning.

Appendix A
Chronology of Domestic Uranium Supply and Demand, 1880–1976

1880	Uranium and vanadium ores are discovered in the Uravan Mineral Belt in Colorado and Utah.
1898	Marie and Pierre Curie discover radium in the uranium ores mined at Joachimstal, Czechoslovakia.
1911–1923	The United States is the world's leading producer of radium, using uranium ores mined from the Colorado Plateau.
1913	Uranium deposits are discovered at Shinkolobwe in the Belgian Congo (now the Republic of Zaire).
25 Feb. 1920	The Mineral Leasing Act is enacted and takes effect immediately. Coal, oil, and gas must be leased, but uranium remains under the Mining Law of 1872.
1930	Discovery of uranium deposits at Great Bear Lake, Canada.
31 July 1939	This is the date chosen by Congress in 1953 to distinguish valid mining locations on Mineral Leasing Act Lands from invalid ones. Mining claims on these lands became valid (as of 1953) if based on locations made on or after this date; earlier claims were invalid. The date was selected to coincide with the beginning of vanadium development on the Colorado Plateau. This was done to promote the production of uranium from these vanadium deposits without allowing a "giveaway" of minerals on older claims.
1940	The U.S. government begins purchasing uranium for the development of the atomic bomb. From 1940 through 1945, all uranium procurement was from three sources: the Shinkolobwe mine in the Belgian Congo, the Eldorado Gold Mines Operation on Great Bear Lake, Canada, and the vanadium mines on the Colorado Plateau.

1944	The United States and the United Kingdom form the Combined Development Trust (CDT) to purchase uranium. The CDT buys uranium from the Belgian Congo. Eldorado Mining and Refining Ltd. becomes the sole agent for uranium purchasing in Canada.
1 Aug. 1946	The Atomic Energy Act of 1946 is enacted. Under section 5(b)(7), all rights to "source material" on public lands are reserved to the U.S. government. Source material includes "uranium, thorium, or any other material which is determined by the Commission, with the approval of the President, to be peculiarly essential to the production of fissionable materials."
1 Jan. 1947	The Atomic Energy Commission, created by the Atomic Energy Act of 1946, assumes all responsibilities previously held by the Manhattan Engineer District. The latter had procured 1,440 tons of uranium from domestic sources and 8,327 tons from foreign sources.
8 Jan. 1947	The Department of the Interior states that all mining claims for uranium are invalid under the Atomic Energy Act of 1946.
1947	The AEC establishes a Division of Raw Materials. All domestic uranium production is a byproduct from four vanadium mills in Colorado.
17 Sept. 1947	The Munitions Board gives the AEC sole responsibility for stockpiling fissionable material. Stockpiles of any other material fall under the board's jurisdiction.
1948	The AEC begins leasing uranium lands on a royalty basis. Also, the AEC acquires a mill at Monticello, Utah and begins rebuilding it in order to use it solely for recovering uranium from ore. The AEC issues Circulars 1 and 2, effective for a 10-year period beginning April 11, 1948. Circular 1 guarantees minimum prices for uranium ores other than carnotite-type or roscoelite-type ores of the Colorado Plateau. Circular 2 provides a $10,000 bonus for the delivery of the initial 20 tons of ore (2,000 ppm or higher) from a mine.

25 Mar. 1948	The AEC begins to withdraw lands in the public domain. Exploration on withdrawn lands is conducted by AEC.
9 Apr. 1948	The AEC issues Circular 3, which guarantees minimum prices for uranium-bearing carnotite-type or roscoelite-type ores of the Colorado Plateau. Terms cover a 3-year period beginning April 11, 1948. Ore deliveries must be made to the AEC's mill in Monticello, Utah.
11 Apr. 1948	Circulars 1, 2, and 3 take effect.
1 June 1948	Initial date of period covered by Circular 4.
15 June 1948	The AEC issues Circular 4, which establishes additional allowances (subsidies) for carnotite-type or roscoelite-type ores of the Colorado Plateau for the period June 1, 1948 to July 1, 1949.
1949	The AEC and USGS issue a booklet entitled "Prospecting for Uranium." According to this booklet, mining claims for uranium and thorium are valid.
1 Feb. 1949	Initial date of period covered by Circular 5.
7 Feb. 1949	The AEC issues Circular 5, guaranteeing minimum prices for carnotite-type or roscoelite-type ores of the Colorado Plateau for the period February 1, 1949-June 30, 1954. Circular 3, which was to expire on April 11, 1951, is superseded. (Circular 5 was later revised, and the expiration date was extended to March 31, 1962.)
1 July 1949	Expiration date of terms of Circular 4.
Sept. 1949	The AEC-owned uranium mill at Monticello, Utah begins operation.
Dec. 1950	The AEC initiates a program to recover uranium from phosphate rock.
12 Feb. 1951	At the request of the AEC, the Secretary of the Interior withdraws a portion of the public lands and federally owned mineral rights in Colorado to permit exploration by the AEC. This withdrawal is accomplished by Public Land Order 698.

26 Feb. 1951	The AEC issues Circular 5, Revised, which covers the period March 1, 1951 to March 31, 1962.
1 Mar. 1951	Circular 5, Revised, goes into effect; period covered by Circular 6 begins also.
Mar. 1951	Jesse Johnson becomes the director of the AEC's Division of Raw Materials.
15 June 1951	A letter written by Senator McMahon leads to the establishment of the Raw Materials Subcommittee of the Joint Committee on Atomic Energy.
27 June 1951	The AEC issues Circular 6, which provides a "bonus for initial production of uranium ores from new domestic mines." The bonus is computed according to the uranium content of the ore. Terms are to be in effect from March 1, 1951 to February 28, 1957 (the latter date was subsequently changed to March 31, 1960).
1951	In South Africa, construction begins on mills designed to recover uranium from gold mine tailings (250 ppm).
June 1952	The Raw Materials Subcommittee of the Joint Committee on Atomic Energy issues its first report.
31 Dec. 1952	Last day on which mineral discovery must be made on mining locations on Mineral Leasing Act Lands in order to qualify for validation under Public Law 250.
1953	The AEC constructs a pilot mill at Grand Junction, Colorado in order to develop improved techniques of uranium recovery from ore.
12 Aug. 1953	Public Law 250 is enacted to validate claims on Mineral Leasing Act Lands. The period from July 31, 1939 to December 31, 1952 is covered by this legislation.
9 Oct. 1953	President Eisenhower delivers his "Atoms for Peace" speech at the United Nations. The program of nuclear exports which followed this speech led to an increase in the demand for uranium.

10 Dec. 1953	Deadline for filing a notice, under Public Law 250, of discoveries made on or before December 31, 1952. Without such notice, claims were not validated by this law.
10 Feb. 1954	The AEC announces that it will use its leasing powers to protect claims made on Mineral Leasing Act Lands after December 31, 1952 and before February 10, 1954. Circular 7 appears in the *Federal Register*.
1 Mar. 1954	Castle Bravo, the first H-bomb, is detonated. The successful development of fusion bombs tends to reduce the growth rate of military demand for uranium.
17–18 May 1954	The Subcommittee on Public Lands of the Senate Committee on Interior and Insular Affairs holds hearing on "multiple mineral use of public lands" (P.L. 585).
20–21 May 1954	The Subcommittee on Mines and Mining of the House Committee on Interior and Insular Affairs holds hearings on "multiple use of mineral lands" (P.L. 585).
19 July 1954	The AEC makes its last withdrawal of lands from location under the mining laws.
13 Aug. 1954	Enactment of the Multiple Mineral Development Act of 1954 (P.L. 585), which validates locations made on Mineral Leasing Act lands after December 31, 1952 and before February 10, 1954. Locations made from February 10 to August 13 are invalid.
30 Aug. 1954	Enactment of Atomic Energy Act of 1954 (42 U.S.C. 2206). Private ownership and operation of nuclear reactors becomes possible.
8 Nov. 1954	The AEC cancels Circular 7, effective December 12, 1954.
11 Dec. 1954	Deadline for making mineral discoveries on locations which can qualify for validation under Public Law 585. Without discovery, mineral locations are invalid.
12 Dec. 1954	Cancellation of Circular 7 is effective.

Apr. 1955   Uranium deposits are discovered at Ambrosia Lake. When it becomes clear that large reserves are present, the AEC's uranium shortage is essentially over. The AEC decides not to sign any new contracts for uranium recovery from phosphate.

21 Dec. 1955   The AEC announces that competitive bidding on its leases will begin on July 14, 1956.

24 May 1956   The AEC announces that it will buy uranium concentrates at a price of $8 per pound from April 1, 1962 to December 31, 1966. Uranium ore purchases are due to expire on March 31, 1962. Circular 6 is revised so as to extend its expiration date to March 31, 1960.

29 June 1956   An act to temporarily postpone assessment work on claims for uranium-bearing lignites is signed (70 Stat. 438).

14 July 1956   Circular 8 is published in the *Federal Register*. It describes regulations for uranium leasing on lands controlled by the AEC. Competitive bidding is required on a cash bonus basis.

1956   The Uranium Institute of America is established.

5 Mar. 1957   Circular 9 is published in the *Federal Register*. It describes regulations for uranium prospecting permits and mining leases on lands administered by federal agencies which do not have the authority to lease such lands.

Apr. 1957   The AEC begins awarding contracts for uranium procurement for the April 1, 1962 to December 31, 1966 perriod.

29 June 1957   Another act regarding lignites is signed (71 Stat. 226).

28 Oct. 1957   Jesse Johnson, the director of the AEC's Division of Raw Materials, announces at an Atomic Industrial Forum conference that the AEC will limit mill construction and expansion, since a shortage no longer exists.

Nov. 1957	Deadline for developing reserves from which concentrate can be sold to the AEC during the April 2, 1958 to March 31, 1962 period.
1958	The AEC closes its mill at Grand Junction, Colorado.
19, 24, 25 Feb. 1958	The Joint Committee on Atomic Energy holds hearings on "problems of the uranium mining and milling industry." Reaction to Jesse Johnson's statement is bitter.
2 Apr. 1958	The AEC announces that it will limit its purchases of uranium concentrate, up to March 31, 1962, to that which is produced from reserves developed prior to November 1, 1957.
11 Apr. 1958	Expiration date of Circulars 1 and 2.
8 May 1958	The AEC announces that it will permit U.S. companies to sell uranium concentrates to foreign and domestic buyers, subject to AEC authorization. Since the AEC owns all special nuclear material (plutonium and uranium enriched in the isotope 233 or in the isotope 235) in domestic reactors, there is no demand for uranium among electric utilities. The first transaction under the new AEC regulation is the sale of U_3O_8 by the Uranium Reduction Co. to the Davison Chemical Division of W.R. Grace & Co.
June 1958	The AEC issues an Ore Reserve Manual describing the agency's method of estimating reserves.
19 Aug. 1958	The Atomic Energy Act of 1954 is amended by Public Law 85-861 to remove the reservation to the United States of source material on federal lands.
24 Nov. 1958	The AEC announces that it will limit its uranium concentrate purchases during the April 1, 1962 to December 31, 1966 period to allocation of production from reserves existing on November 24, 1958 and identified by drilling.
9 May 1959	The AEC again revises Circular 6, in order to bring about an orderly termination of production bonuses. Closing of the Monticello, Utah mill is scheduled for January 1, 1960.

18 May 1959	The AEC announces that companies seeking contracts for the April 1, 1962 to December 31, 1966 period must submit data on their 1958 reserves to the AEC no later than July 31, 1959 (this deadline was subsequently extended to October 1, 1959).
1 July 1959	All AEC purchases of uranium from Canada are consolidated under one contract, effective July 1.
15 July 1959	The AEC offers to help companies meet the July 31 deadline for reserve data required for U_3O_8 contracts in the April 1, 1962 to December 31, 1966 perriod.
31 July 1959	Initial deadline for submittal of data on November 24, 1958 reserves.
14 Aug. 1959	The AEC announces that it will extend the deadline to October 1, 1959.
1 Oct. 1959	Final deadline for reserve data.
15 Jan. 1960	The AEC shuts down its mill at Monticello, Utah.
31 Mar. 1960	Expiration date of Circular 6, Revised.
7 June 1960	The General Accounting Office issues a report on its review of the AEC contract with the Anaconda Company.
14 June 1960	The General Accounting Office issues a report on its review of the AEC contract with Union Carbide Nuclear Company.
1960	The International Resources Corp. proposes to build a mill at Bowman, N.D. to recover uranium from lignite. The mill would have a capacity of 200 short tons of lignite per day, through March 31, 1962 and 600 tons per day thereafter. The contract with the AEC is not executed, however.
Dec. 1960	The AEC receives its last delivery of uranium from Union Miniere du Haut Katanga, the company operating in the Congo.
1 Jan. 1961	New contracts for uranium from South Africa go into effect. The Combined Development Agency contract is replaced by separate U.S. and U.K. contracts.

19 Jan. 1961	The General Accounting Office issues a report on its review of the AEC contract with Western Nuclear Corporation.
31 Jan. 1961	The General Accounting Office issues a report on its review of the AEC contract with Lucky Mc Uranium Corporation.
21 Feb. 1961	The AEC announces that it will sign some contracts for uranium concentrate delivered in the April 1, 1962 to December 31, 1966 period but not produced from November 24, 1958 reserves. For "hardship" cases with low reserves in the Uravan Mineral Belt, contracts may be assigned according to historical production levels.
31 Mar. 1961	The AEC's last contracts for vanadium oxide (V_2O_5) purchases expire.
1 July 1961	The AEC's last contracts for uranium production from phosphate expire.
Aug. 1961	The AEC delivers a summary report on raw materials to the Joint Committee on Atomic Energy.
15–16 Nov. 1961	The Subcommittee on Raw Materials of the Joint Committee on Atomic Energy holds hearings on the AEC raw materials program. Gordon Weller, representing the Uranium Institute of America, charges that the AEC awarded large contracts to Kerr-McGee Oil Industries, Inc. because Robert S. Kerr, chairman of the board of Kerr-McGee, was a ranking member of both the Senate Finance Committee and the Senate Public Works Committee.
Dec. 1961	The AEC receives its last delivery of uranium from Radium Hill, Australia, with the expiration of a Combined Development Agency contract.
Mar. 1962	The General Accounting Office issues its "Review of Selected Aspects of the Domestic Uranium Procurement Program." The AEC is criticized for setting a fixed price on April 1, 1962 to December 31, 1966 deliveries instead of writing cost-plus contracts for each firm.

31 Mar. 1962	Circular 5, Revised, expires, and so the AEC stops buying ore from domestic producers. All mining ceases on lands leased by the AEC.
Apr. 1962	The AEC receives its last delivery of uranium from Urgeirica, Portugal, with the expiration of a Combined Development Agency contract.
May 1962	The Atomic Industrial Forum (AIF) issues a report on "The Domestic Uranium Raw Materials Industry." This report by the AIF's Committee on Mining and Milling warns that nearly all U.S. uranium producers will go out of business unless the AEC extends its purchase program. (Subsequent events show that the AIF seriously underestimated the uranium demand associated with the civilian nuclear power industry.)
18–19 June 1962	The Subcommittee on Raw Materials of the Joint Committee on Atomic Energy holds hearings on the AEC uranium procurement program.
July 1962	The AEC opens a temporary field office at Bowman, N.D. in order to allocate the amounts of uranium concentrate which may be produced from lignite by December 31, 1966. (Subsequently the idea of recovering uranium from lignite, after burning the lignite *in situ*, was abandoned.)
Nov. 1962	The AEC announces its "stretchout" program—an offer to extend the deadline for uranium concentrate purchases beyond December 31, 1966. Uranium delivered in 1967 and 1968 must be produced from reserves which were proven by November 24, 1958 (as before), but uranium delivered in 1969 and 1970 can equal the amount delivered in 1967 and 1968, regardless of when it entered the reserve base.
Jan. 1963	The AEC receives its last delivery of uranium from Australia, with the expiration of the contract for production at Rum Jungle. This is the last delivery to the United States under a Combined Development Agency contract.
30, 31 July, 1 Aug. 1963	The Joint Committee on Atomic Energy holds hearings on private ownership of special nuclear materials.

17 Feb. 1964 Jersey Central Power and Light Co. issues its "Economic Analysis for Oyster Creek Nuclear Electric Generation Station." Generating costs are estimated to be very close to those which would be incurred by a coal plant of similar capacity. The achievable power level of the plant remains to be determined through operating experience, so costs per kilowatt-hour are uncertain. Despite the fact that the AEC owns the fuel, Jersey Central P and L announces that commercial nuclear power has arrived.

9, 10, 11, 15, 25 June 1964 The Joint Committee on Atomic Energy holds additional hearings on private ownership of special nuclear materials.

26 Aug. 1964 The Private Ownership of Special Nuclear Materials Act is approved. *Special nuclear materials* are uranium enriched in the isotopes 235 or 233 and plutonium. Thus this act allows private ownership or nuclear fuel, and opens up a market for commercial purchasing of uranium concentrates for use in domestic power reactors. Under this law, Section 161v of the Atomic Energy Act of 1954 is amended to require that the AEC shall not enrich foreign uranium for domestic use, to the extent necessary to ensure the maintenance of a viable domestic uranium industry. Since the AEC is the only owner of enrichment plants, the effect of this provision is to prevent U.S. power companies from buying foreign uranium. A schedule for phasing in the enrichment of foreign uranium for domestic use is not issued by the AEC until October 25, 1974.

1964 Kermac Nuclear Fuels Corp. starts a plant at Bowman, N.D. to burn 225 tons of lignite per day and ship 75 tons of uranium-bearing ash per day to a uranium mill. Union Carbide starts a similar plant at Belfield, N.D.

1965 The Organization for Economic Cooperation and Development, together with the European Nuclear Energy Agency, issues a report on "World Uranium and Thorium Resources."

Sept. 1965 The Susquehanna Corp. announces a contract to supply 475,000 lb U_3O_8 to Allegmeine Elektrizitaets

Gesellschaft of West Germany. This is said to be the largest private contract to date among U.S. producers.

1966   Private purchases of domestically produced uranium for use in domestic reactors are initiated.

23 Dec. 1966   The AEC publishes a notice in the *Federal Register* stating that the enrichment of foreign uranium for domestic use would threaten the viability of the domestic uranium industry and therefore must be prohibited until such time as it poses less of a threat.

31 Dec. 1966   This is the deadline announced on May 24, 1956 for delivery of uranium concentrates by domestic producers. Deliveries continue, however, under the "stretchout" program. Foreign producers are not afforded such protection; the AEC stops importing uranium upon the expiration of contracts with Canadian and South African producers.

9, 10, 23 May,
6, 7, 8, 9 June,
26, 27 July,
8, 10 Aug. 1967   The Joint Committee on Atomic Energy holds hearings on the radiation exposure of uranium miners. There exists a controversy over the need to establish standards which limit the inhalation of radon.

1967   The mining of uranium-bearing lignite in Harding County, S.D. ceases.

28 Dec. 1967   Philip Sporn, retired president and director of American Electric Power Co., sends an analysis of nuclear power economics to the Joint Committee on Atomic Energy, pursuant to the chairman's request. Sporn observes that nuclear power costs for plants on order have risen since 1965. The rise in generation costs threatens the long-term growth of the utility industry, in Sporn's view.

1 July 1968   United Nuclear Corp. files suit against Combustion Engineering, Inc. This is the uranium industry's first antitrust case. United Nuclear, a uranium producer trying to sell nuclear fuel reloads to utilities, opposed the acquisition of 21 percent of its stock by Combustion Engineering, a reactor manufacturer. On July 3, 1969, the court decided in favor of United Nuclear.

1968   The Uranium Institute of America, the producers' association, ceases effective operation. The AEC issues its first annual report on "Statistical Data of the Uranium Industry." Activities of the AEC's Grand Junction Office allow uranium mining companies to learn about industrywide trends without setting up a trade association.

Dec. 1968   Arthur D. Little, Inc. completes a report on "Competition in the Nuclear Power Supply Industry," under a contract issued by the AEC and the Department of Justice. The industrial consulting firm states that "all factors point to a continued viable and competitive uranium mining and milling industry after government purchases are terminated in 1970, as is planned." No attention is given to the possibility of an international cartel.

1 Jan. 1969   The AEC is permitted to begin enriching privately owned uranium under the terms of the Private Ownership of Special Nuclear Materials Act. Enrichment services are provided under "requirements" contracts, that is, contracts which allow utilities to state the amount of separative work they require on a short-term basis, subject to minimum and maximum limits. Both the transaction tails assay and operating tails assay are set at 0.20 percent.

17–18 Mar. 1969   The Joint Committee on Atomic Energy holds hearings on "Radiation Standards for Uranium Mining."

9 Apr. 1969   The AEC awards an operating license for the Oyster Creek reactor. This marks the start of operation of commercial nuclear power plants.

1 Dec. 1969   At a speech before the Atomic Industrial Forum, AEC Commissioner Wilfred E. Johnson suggests a partial lifting of the restriction on enrichment of foreign uranium for domestic use, beginning in 1973, and the sale by the AEC of its surplus uranium stocks acquired during the "stretchout" program.

10 Nov. 1970   A proposed modification of Circular 8 is published in the *Federal Register*.

31 Dec. 1970	AEC procurement of uranium ends. (A few deliveries are allowed to be postponed until 1971, however.) In addition, this is the deadline set by the Private Ownership of Special Nuclear Materials Act of 1964 for issuance of nuclear fuel leases by the AEC.
1 July 1971	AEC increases the operating tails assay from 0.20 percent to 0.30 percent, while maintaining the transaction tails assay at 0.20 percent. This marks the de facto introduction of a "split tails" policy.
23 July 1971	The U.S. Court of Appeals for the District of Columbia Circuit issues a decision, in *Calvert Cliffs Coordinating Committee* v. *AEC*, requiring that environmental impact statements be written and reviewed in licensing proceedings for nuclear power plants. Four years elapse before this decision has a direct impact on uranium mills.
17 Aug. 1971	James R. Schlesinger becomes chairman of the AEC, succeeding Glenn T. Seaborg; Schlesinger is a newcomer and outsider. Seaborg, who retires after a 10-year term as chairman, represents the Manhattan Project scientists, the "old guard."
13 Oct. 1971	The AEC proposes to dispose of its stockpile of surplus uranium by selling it to domestic and foreign customers. Public comment is requested. The uranium producers object strenuously to the plan, since it would reduce their sales.
28–29 Oct. 1971	The Joint Committee on Atomic Energy holds hearings on "Use of Uranium Mill Tailings for Construction Purposes."
Feb. 1972	Representatives of Australia, Canada, France, and South Africa meet in Paris to discuss the problem of low prices for uranium in the world market. Australia, Canada, and South Africa are the three major producing countries outside the United States and the Communist countries. A second meeting in Paris is held in March, and a third meeting is held in Johannesburg in June. Cartel action to set quotas and minimum prices is considered.

Mar. 1972	The AEC rescinds its proposal to sell uranium and adopts a plan to operate its enrichment plants under a "split tails" system. The actual tails assay is to be higher than that stated in enrichment contracts. As a result, feed requirements are increased while electricity and separative work requirements are decreased. This plan represents a way to dispose of the stockpile without cutting into producers' sales.
Apr. 1972	The AEC issues a report on its "Nuclear Industry Fuel Supply Survey," providing data on current supply and demand. Subsequently, an annual "Survey of U.S. Uranium Marketing Activity" is published every April.
June 1972	The AEC issues its first annual report on its survey of exploration expenditures and costs per foot of drilling.
Sept. 1972	The AEC issues an environmental impact statement on "Leasing of AEC Controlled Uranium Bearing Lands, Colorado, Utah, New Mexico."
6 Dec. 1972	The AEC publishes "Grand Junction Remedial Action Criteria" in the *Federal Register*. These regulations involve mill tailings used in construction of houses and schools.
6 Feb. 1973	Dixy Lee Ray becomes the seventh (and last) chairman of the AEC, succeeding James Schlesinger.
Mar. 1973	The AEC begins the National Uranium Resource Evaluation—a program to estimate the quantity of "potential resources" of uranium which are yet to be discovered in the United States.
27 Mar. 1973	At the Atomic Industrial Forum's Uranium Seminar, Robert Nininger of the AEC warns of the inconsistency between current uranium prices and the need for a major exploration program to begin.
Apr. 1973	The AEC issues reports on "Nuclear Fuel Supply" and "Nuclear Fuel Resources and Requirements," stressing the need to expand reserves in order to keep pace with projected uranium demand.
9 May 1973	The AEC changes its enrichment contracting procedures. New contracts must be "fixed-commitment

contracts," that is, utilities must fix the amount of enrichment service while they are committed to purchase on a long-term planning basis. A December 31, 1973 deadline is set for execution of contracts for initial delivery of enriched uranium before July 1, 1978. A June 30, 1974 deadline is set for contracts for initial delivery of enriched uranium between July 1, 1978 and June 30, 1982.

30 June 1973   All AEC leases of nuclear fuel for commercial and industrial use expire in accordance with the deadline set by the Private Ownership of Special Nuclear Materials Act of 1964.

10 Aug. 1973   A revised version of Circular 8 is published in the *Federal Register*. It contains the regulations for uranium leases on lands controlled by AEC.

19 Sept. 1973   In an AEC news release, the Grand Junction Office announces a uranium leasing program. By the end of 1973 the AEC has issued 230 invitations for bids.

16 Oct. 1973   The posted price for Saudi Arabian light oil (34°, fob Ras Tanura) rises from $2.90 per barrel to $5.12 per barrel. OPEC wields its power.

27 Nov. 1973   The AEC publishes a notice in the *Federal Register* about its proposal to lift the restriction on enrichment of foreign uranium for domestic use.

31 Dec. 1973   Deadline for execution of enrichment contracts specifying initial delivery before July 1, 1978.

1 Jan. 1974   The posted price for Saudi Arabian light oil rises from $5.12 per barrel to $11.65 per barrel. Subsequently, the U.S. government responds by promoting nuclear power as a substitute for imported oil.

12 Mar. 1974   The Joint Committee on Atomic Energy holds hearings on "Uranium Mill Tailings in the State of Utah."

May 1974   The General Accounting Office issues its report on "Modernization of 1872 Mining Law Needed to Encourage Domestic Mineral Production, Protect the Environment, and Improve Public Land Manage-

	ment.'' The GAO recommends that Congress enact legislation to ''establish a leasing system for extracting minerals from public lands.''
9 June 1974	Contrary to its May 1973 announcement, the AEC suspends the signing of long-term enrichment contracts for all countries except Egypt, Israel, and Iran, for which President Nixon has made commitments (conditioned upon nonproliferation agreements). In effect, the June 30 deadline is advanced 3 weeks. This action is taken because the demand for AEC's enrichment services exceeds the supply projected for the future.
30 June 1974	Deadline for execution of enrichment contracts specifying initial delivery of enriched uranium between July 1, 1978 and June 30, 1982.
2 July 1974	The AEC stops accepting contracts for long-term enrichment services. It will not even examine new contracts to consider signing them.
6 Aug. 1974	The AEC decides to resume signing of contracts already received, under a scheme in which forty-five of the foreign requests must be conditioned upon regulatory approval of plutonium recycle in light-water reactors.
14 Aug. 1974	The AEC finishes signing enrichment contracts for domestic reactors.
17–18 Sept. 1974	The Joint Committee on Atomic Energy holds hearings on the AEC's ''Proposed Modification of Restrictions on Enrichment of Foreign Uranium for Domestic Use.'' The modifications are announced on October 25.
Oct. 1974	The AEC finishes signing all the standard contracts for enrichment services which have been pending since June. The ''conditional'' contracts remain to be resolved.
11 Oct. 1974	The Energy Reorganization Act of 1974 is approved. Under this act the AEC is to be replaced by the Energy Research and Development Administration (ERDA) and the Nuclear Regulatory Commission (NRC).

25 Oct. 1974   The AEC announces its new policy to lift restrictions on enrichment of foreign uranium for domestic use. The amount of foreign uranium feed which any customer delivers must not exceed 10 percent of his deliveries in 1977, 15 percent in 1978, 20 percent in 1979, 30 percent in 1980, 40 percent in 1981, 60 percent in 1982, and 80 percent in 1983. No restrictions are to be in effect after 1983.

19 Jan. 1975   The AEC is replaced by ERDA and NRC. Also in January, the Joint Committee on Atomic Energy begins the 94th Congress without four of its senior members—George Aiken, Wallace Bennett, Chet Holifield, and Craig Hosmer. Thus weakened, the JCAE can survive only 2 more years.

28 Mar. 1975   The Natural Resources Defense Council petitions the NRC for a generic environmental statement on uranium milling. Concern about radon gas released from mill tailings is the focus of this petition.

21 May 1975   The General Accounting Office issues a report on "Controlling the Radiation Hazard from Uranium Mill Tailings."

19 June 1975   ERDA decides that "fixed commitment" contracts do not have to be fixed after all. Holders of these contracts are allowed to adjust or terminate them by August 18. (This represents a 2-month "open season.") After August 18, customers may terminate ERDA contracts only in order to contract with domestic private enrichment companies. ERDA announces a transaction tails assay (for contract purposes, not for operating requirements) of 0.20 percent through September 30, 1977, and proposes an increase to 0.25 percent through September 30, 1979, 0.275 percent through September 30, 1981, and 0.30 percent thereafter.

26 June 1975   President Ford proposes a program to Congress which would promote a privately owned domestic enrichment industry. The Nuclear Fuel Assurance Act does not gain widespread approval, however.

1 July 1975   AEC lowers the operating tails assay from 0.30 per-

cent to 0.25 percent, while maintaining the transaction tails assay at 0.20 percent.

8 Sept. 1975   Westinghouse Electric Corp. sends notices to nineteen U.S. utilities, saying that it does not intend to meet its contract commitments for delivery of uranium concentrate because the price of uranium has risen sharply.

Oct. 1975   Utilities respond by suing Westinghouse.

May 1976   Homestake Mining Co. sues the Washington Public Power Supply System in order to regain 300,000 pounds of uranium which should never have been delivered to WPPS, according to Homestake, because it was not needed to operate the Hanford 2 reactor.

June 1976   The NRC announces that it will prepare a generic environmental impact statement on uranium milling.

2 July 1976   The General Accounting Office issues a report on "Certain Actions That Can Be Taken to Help Improve This Nation's Uranium Picture." Concern about uranium exports is expressed.

Aug. 1976   Friends of the Earth Australia gives the U.S. Department of Justice copies of documents from the files of Mary Kathleen Uranium Ltd. which provide evidence that meetings in Paris in February 1972 led to international price fixing.

1 Oct. 1976   The Senate votes down the Nuclear Fuel Assurance Act. Growth in U.S. enrichment capacity must therefore be accomplished by additions to ERDA's Portsmouth, Ohio plant rather than by private domestic plants.

14 Oct. 1976   The Federal Trade Commission files suit against Atlantic Richfield Co. for its attempt to acquire the Anaconda Company. The FTC alleges that the merger would restrict competition in both the uranium oxide and copper industries.

15 Oct. 1976   Westinghouse Electric Corp. files an antitrust suit against 30 uranium producers, alleging price fixing and seeking treble damages.

20 Oct. 1976   John B. Martin of the NRC announces at ERDA's annual uranium industry seminar that licensing delays of a year or more may result from the preparation and review of environmental impact statements for uranium mills.

Appendix B
Uranium Milling Companies in the United States

These companies operate existing mills or have made commitments to construct mills. Subsidiaries and joint ventures are listed. Subsidiaries are wholly owned unless otherwise noted.

1. **Domestic Companies Involved in Uranium Exploration, Mining, and Milling Outside the United States (see appendix C)**

   Continental Oil Company: joint venture with Pioneer Natural Gas (63.7 percent Continental, 36.3 percent Pioneer)
   Exxon Corporation
     Exxon Nuclear Company
   Federal Resources Corporation: joint venture with American Nuclear Corporation, called Federal-American Partners (60 percent Federal, 40 percent American)
     Allied Nuclear Corporation (86.9 percent owned)
   Getty Oil Corporation: joint venture with Kerr-McGee Corporation (q.v.) and Skelly Oil Corporation
     Mission Corporation (87.93 percent owned)
     Skelly Oil Corp. (72.53 percent owned by Mission and 3.56 percent owned by Getty): joint venture with Kerr-McGee Corporation (q.v.) and Getty Oil Corporation
   Kerr-McGee Corporation
     Kerr-McGee Nuclear Corporation: joint venture with Getty Oil Corporation and Skelly Oil Corporation, called Petrotomics Company (50 percent Kerr-McGee, 33.3 percent Getty, 16.7 percent Skelly)
   Phelps Dodge Corporation
     Western Nuclear, Incorporated
   Westinghouse Electric Corporation
     Wyoming Mineral Corporation

2. **Foreign Companies**

United Kingdom

   Rio Tinto Zinc
     Preston Mines, Ltd. (80.9 percent owned)

Rio Algom, Ltd. (43.9 percent owned by Preston Mines and 51.3 percent owned by Rio Tinto Zinc)
Rio Algom Corporation

3. Other Diversified Companies

Anaconda Company
General Electric Company
   Lucky Mc Uranium Company (formerly owned by Utah International, Incorporated, which was acquired by GE in 1976)
Homestake Mining Company: joint venture with United Nuclear Corporation (q.v.)
Mobil Oil Corporation
Newmont Mining Corporation
   Dawn Mining (51 percent owned)
Pioneer Natural Gas: joint venture with Continental Oil Company (q.v.)
Reserve Oil and Minerals: joint venture with Standard Oil Company of Ohio (50 percent Sohio, 50 percent Reserve)
Standard Oil Company of Ohio: joint venture with Reserve Oil and Minerals (q.v.)
Standard Oil Company of California
Union Carbide Corporation
Union Oil Company of California
Union Pacific Railroad Corporation
   Rocky Mountain Energy Company
United States Steel Corporation

4. Companies for which Uranium is a Primary Line of Business

American Nuclear Corporation: joint venture with Federal Resources Corporation (q.v.)
Atlas Corporation
United Nuclear Corporation: joint venture with Homestake Mining Company, called United Nuclear–Homestake Partners (70 percent United Nuclear, 30 percent Homestake)

5. Utility Companies

Commonwealth Edison
   Cotter Corporation (acquired in 1974)

Appendix C
Companies Involved in Uranium Exploration, Mining, and Milling Outside the United States

There is no basic reference work which describes the various corporations involved in uranium production worldwide. The following list has been compiled from news articles and other sources. We hope that it will provide a reasonably accurate picture of the structure of the world market, but no attempt has been made to assemble a complete list of exploration programs. Subsidiaries and joint ventures are listed. Subsidiaries are wholly owned unless otherwise noted.

1. Privately Owned Industrial Corporations

Australia

Note: None of these corporations, except Western Mining Corporation, Ltd., is listed in *Moody's Industrial Manual, 1976*. Therefore, some of them may be controlled by larger corporations which we have been unable to identify. All joint ventures listed here are located in Australia.

Central Coast Exploration N.L.: joint venture with Getty Oil Corporation

Central Pacific Minerals N.L.: joint venture with Urangesellschaft (q.v.), Ente Nazionale Idrocarburi, Magellan Petroleum Australia Ltd., and others

Electrolytic Zinc Company of Australasia Ltd.: joint venture with Peko-Wallsend Ltd., called Ranger Uranium Mines Pty. Ltd.

Oilmin N.L.: joint venture with Phelps Dodge Corporation (q.v.), Petromin N.L., and Transoil N.L.

Pancontinental Mining Ltd.: joint venture with Getty Oil Corporation, at Jabiluka (65 percent Pancontinental, 35 percent Getty)

Peko-Wallsend Ltd.: joint venture with Electrolytic Zinc Company of Australasia Ltd. (q.v.); joint venture with Western Mining Corporation Ltd., called Geopeko

Petromin N.L.: joint venture with Phelps Dodge Corporation (q.v.), Oilmin N.L., and Transoil N.L.

Queensland Mines Ltd. (controlled by Kathleen Investments Ltd. in 1971): joint venture with Westinghouse Electric Corporation—proposed in 1973 but not approved by the Australian government

Transoil N.L.: joint venture with Phelps Dodge Corporation (q.v.), Oilmin N.L., and Petromin N.L.

Vam Ltd.: joint venture with Delhi International Oil Corporation and Westinghouse Electric Corporation

Western Mining Corporation, Ltd.: joint venture with Peko-Wallsend Ltd. (q.v.)

Canada

Denison Mines, Ltd.
  Stanrock Uranium Mines Ltd. (acquired in 1973): joint venture in Ontario, Canada, with Uranium Exploration Company (see Mitsui & Co.)
Noranda Mines Ltd.
  Noranda Australia Ltd.
  Kerr Addison Mines Ltd. (34 percent owned)
    Keradamex
    Agnew Lake Mines Ltd.
Numac Oil and Gas Ltd.: joint venture in Saskatchewan, Canada, with Imperial Oil Ltd. (see Exxon Corporation)

France

Compagnie Francaise des Petroles (35 percent owned by the government of France)
  TOTAL Compagnie Miniere et Nucleaire
    Rossing Uranium Ltd. (10 percent owned; see Rio Tinto Zinc) joint venture with Pechiney Ugine Kuhlmann, called Minatome; joint ventures in Niger and Gabon with the Commissariat a l'Energie Atomique; joint venture in Mauretania with Marubeni Corporation and Nippon Mining
Imetal
  Societe Miniere et Metallurgique de Penarroya
  Compagnie de Mokta
  Compagnie des Mines d'Uranium de Franceville
  Compagnie Francaise des Minerais d'Uranium: joint venture in Saskatchewan, Canada with Pechiney Ugine Kuhlmann and the Com-

missariat a l'Energie Atomique, called Amok Ltd. (45 percent owned
by Mokta and CFMU); joint venture in Niger, with Pechiney Ugine
Kuhlmann, the Commissariat a l'Energie Atomique (q.v.), the government of Niger, Ente Nazionale Idrocarburi, and Urangesellschaft

   Uranex (49.5 percent owned in 1971; established in 1969 by Mokta,
   CFMU, Pechiney Ugine Kuhlmann, and the Commissariat a
   l'Energie Atomique)
Pechiney Ugine Kuhlmann: joint venture with Compagnie Francaise
des Petroles (q.v.); joint venture in Saskatchewan, Canada, with
Compagnie de Mokta and Compagnie Francaise des Minerais
d'Uranium (see Imetal) and the Commissariat a l'Energie Atomique;
joint venture in Niger with Compagnie de Mokta, Compagnie Francaise des Minerais d'Uranium, the Commissariat a l'Energie
Atomique (q.v.), the government of Niger, Ente Nazionale Idrocarburi, and Urangesellschaft; joint venture in the United States, with
Mitsubishi Corporation

Japan

   Arabian Oil Company
   International Resources Development Company: joint venture in
      Niger with the government of Niger
   Marubeni Corporation: joint venture in Mauretania with Minatome
   (see Compagnie Francaise des Petroles) and Nippon Mining
      Brinco Ltd. (8 percent owned by Marubeni Corporation and Fuji
      Bank; see Rio Tinto Zinc)
   Mitsubishi Corporation: joint venture in the United States with
   Pechiney Ugine Kuhlmann; joint venture in Australia with Mitsubishi
   Metal Corporation, called Tahei Uranium Exploration; joint venture
   in Australia with Westinghouse—proposed in 1974
   Mitsui and Company: joint venture with Mitsui Mining and Smelting
   Company, called Uranium Exploration Company (63.67 percent Mit-
   sui, 36.33 percent Mitsui Mining and Smelting); joint venture in On-
   tario, Canada, with Denison Mines, Ltd.
   Nippon Mining: joint venture in Mauretania with Minatome (see Com-
   pagnie Francaise des Petroles) and Marubeni Corporation
   Overseas Uranium Resources Development Company: joint venture in
   Niger with the Commissariat a l'Energie Atomique (q.v.) and the
   government of Niger; joint venture in the Somali Republic with Ente
   Nazionale Idrocarburi

United Kingdom

Rio Tinto Zinc
  Preston Mines, Ltd. (80.9 percent owned)
    Rio Algom Ltd. (43.9 percent owned by Preston Mines and 51.3 percent owned by Rio Tinto Zinc)
  Rio Algom Corporation
  Conzinc Riotinto of Australia Ltd. (80.6 percent owned)
    Northern Territory Enterprises Pty. Ltd.
  Mary Kathleen Uranium Ltd. (41.1 percent owned)
  RTZ Services
  Palabora Mining Company, Ltd. (38.9 percent owned)
  Rossing Uranium Ltd. (53.6 percent owned): originally formed as a joint venture among Rio Tinto Zinc, General Mining and Finance Corporation Ltd., and Industrial Development Corporation of South Africa
  Brinco Ltd. (62.7 percent owned): originally formed in 1968 when Rio Algom Ltd. sold British Newfoundland Corporation Ltd. to Rio Tinto Zinc and Bethlehem Steel Corporation, each holding 50 percent
    Brinex: joint venture in Labrador, Canada, with Metallgesellschaft (see Urangesellschaft; 60 percent Brinex, 40 percent Metallgesellschaft)

United States

Bethlehem Steel Corporation
  Brinco Ltd. (50 percent owned in 1968; see Rio Tinto Zinc)
Continental Oil Company
  Hudson's Bay Oil and Gas Company, Ltd. (53.1 percent owned): attempted to buy 35.7 percent of Denison Mines Ltd. in 1970, but failed to obtain Canadian government approval; joint venture in Niger with the Commissariat a l'Energie Atomique
Delhi Internation Oil Corporation: joint venture in Australia with Westinghouse Electric Corporation and Vam Ltd.
Exxon Corporation
  Imperial Oil Limited (69.52 percent owned): joint venture in Saskatchewan, Canada, with Numac Oil and Gas Ltd.
  Exxon Nuclear Company
  Exxon Nuclear International, Inc.
Federal Resources Corporation
  Can-Fed Resources Corporation
  Madawaska Ltd. (51 percent owned)

Getty Oil Corporation
  Getty Oil Development Company, Ltd.: joint venture in Australia,
    with Pancontinental Mining Ltd. (q.v.); joint venture in Australia,
    with Central Coast Exploration N.L.
Gulf Oil Corporation
  Gulf Minerals Company
  Gulf Minerals Canada Ltd.: joint venture in Saskatchewan, Canada,
    with Uranerz Canada Ltd. and Gulf Oil Canada Ltd., at Rabbit Lake
    (46 percent Gulf Minerals Canada, 49 percent Uranerzbergbau, 5.1
    percent Gulf Oil Canada)
  Gulf Oil Canada Ltd. (68.26 percent owned): joint venture in Sas-
    katchewan, Canada, with Gulf Minerals Canada Ltd. (q.v.) and
    Uranerz Canada Ltd.
Inexco Oil Company
  Inexco Mining Company (Canada) Ltd.: joint venture in Saskatche-
    wan, Canada, with Uranerz Exploration and Mining Co. and Sas-
    katchewan Mining and Development Corporation (q.v.)
Kerr-McGee Corporation
  Kerr-McGee Nuclear Corporation
Magellan Petroleum Corporation
  Magellan Petroleum Australia Ltd.: joint venture in Australia with
    Ente Nazionale Idrocarburi, Urangesellschaft (q.v.), Central
    Pacific Minerals N.L., and others
Phelps Dodge Corporation
  Western Nuclear, Inc.
  Western Uranium Ltd.: joint venture in Australia with Oilmin N.L.,
    Petromin N.L., and Transoil N.L., called South Australian
    Uranium Corporation Pty. Ltd. (50 percent Western, 50 percent
    others)
Westinghouse Electric Corporation
  Wyoming Mineral Corporation: joint venture in Australia with Delhi
    International Oil Corporation and Vam Ltd.
  Westinghouse Nuclear Europe, Inc.: joint venture in Australia with
    Mitsubishi Corporation—proposed in 1974; joint venture in Aus-
    tralia with Queensland Mines Ltd.—proposed in 1973, but not
    approved by the Australian government

West Germany

Rheinische Westfaelisches Elektrizitaetswerke
  C. Deilmann and Rheinische Braunkohlenwerke
  Uranerzberbau G.m.b.H. & Co.: joint venture in Saskatchewan,
    Canada, with Gulf Minerals Canada Ltd. (q.v.) and Gulf Oil Canada
    Ltd.

Uranerz Exploration and Mining Co.: joint venture in Saskatchewan, Canada, with Inexco Oil Company and Saskatchewan Mining
and Development Corporation (q.v.); joint venture in Australia (in
1970) with Urangesellschaft

2. Consortia Established by Privately Owned Corporations

Japan

Power Reactor and Nuclear Fuel Development Corporation (owned by
ten companies and the Japanese government in 1971): joint venture in
Niger with the Commissariat a l'Energie Atomique, the government
of Niger, and Urangesellschaft, called Societe des Mines du Djado (25
percent PRNFDC, 75 percent others)
Tokyo Uranium Development Corporation (owned by five companies
in 1974)

West Germany

Urangesellschaft A. G. (one-third owned by Metallgesellschaft, which
entered a joint venture with Brinex; see Rio Tinto Zinc): joint venture
in Niger with the Commissariat a l'Energie Atomiqe (q.v.), Compagnie de Mokta, Compagnie Francaise des Minerais d'Uranium, the
government of Niger, and Ente Nazionale Idrocarburi; joint venture
in Niger with Power Reactor and Nuclear Fuel Development Corporation (q.v.), the Commissariat a l'Energie Atomique, and the government of Niger; joint venture in Australia with Ente Nazionale
Idrocarburi, Magellan Petroleum Australia Ltd., Central Pacific Minerals N.L., and others, at Ngalia (65 percent Urangesellschaft and
ENI, 10 percent Magellan, 25 percent others); joint venture in Australia (in 1970) with Uranerzberbau; joint ventures in Angola and
Mozambique with the Junta de Energia Nuclear

3. Privately Owned South African Financial Corporations with Investments in South African Gold Mines

This list is intended to cover all South African gold mines (including
those which do not presently recover uranium), together with the
Palabora copper mine in South Africa and the Rossing uranium mine in
South West Africa (Namibia).

Anglo American Corporation of South Africa, Ltd.
    Engelhard Minerals and Chemicals Corporation (30 percent owned)

Charter Consolidated (major share owned; Charter Consolidated owns nearly 10 percent of Rio Tinto Zinc)

Anglo American Gold Investment Company, Ltd. (major share owned)

Johannesburg Consolidated Investment Company, Ltd. (controlled)

AAC is Secretary of the following companies:
Western Holdings Ltd.
Jeanette Gold Mines Ltd. (partially owned)
St. Helena Gold Mines Ltd. (partially owned; see Union Corporation, Ltd.)
Free State Geduld Mines Ltd.
Freddies Consolidated Ltd. (acquired in 1964 by Western Holdings and Free State Geduld Mines, and jointly held)
Welkom Gold Mining Company, Ltd.
President Brand Gold Mining Company, Ltd.
Free State Saaiplaas Gold Mining Company, Ltd.
President Steyn Gold Mining Company, Ltd.

Johannesburg Consolidated Investment Company, Ltd. is Secretary of the following companies:
Western Areas Gold Mining Company, Ltd.
Elsburg Gold Mining Company, Ltd.

Anglo-Transvaal Consolidated Investment Company, Ltd.
Middle Witwatersrand (Western Areas) Ltd. (51 percent owned)

ATCI is Secretary of the following companies:
Hartebeestfontein Gold Mining Company, Ltd.
Zandpan Gold Mining Company, Ltd.
Loraine Gold Mines, Ltd.

Barlow Rand Ltd.
Rand Mines Ltd. is Secretary of the following companies:
Blyvooruitzicht Gold Mining Company, Ltd.
Harmony Gold Mining Company, Ltd.
Merriespruit (Orange Free State) Gold Mining Company, Ltd.
Virginia Orange Free State Gold Mining Company, Ltd.

General Mining and Finance Corporation, Ltd.
West Rand Consolidated Mines, Ltd.
Rossing Uranium Ltd. (partially owned in 1971; see Rio Tinto Zinc)
Union Corporation Ltd. (majority share owned)

GM&F is Secretary of the following company:
Buffelsfontein Gold Mining Company

Union Corporation is Secretary of the following companies:
  Kinross Mines Ltd.
  Leslie Gold Mines Ltd.
  St. Helena Gold Mines Ltd.
  Winkelhaak Mines Ltd.

Gold Fields of South Africa Ltd. is Secretary of the following companies:
  Doornfontein Gold Mining Company, Ltd.
  East Driefontein Gold Mining Company, Ltd.
  Kloof Gold Mining Company, Ltd.
  West Driefontein Gold Mining Company, Ltd.
  Libanon Gold Mining Company, Ltd.
  Venterspost Gold Mining Company, Ltd.

ASA Ltd. (formerly American-South African Investment Company, Ltd.) owns securities in numerous gold mining companies

Industrial Development Corporation of South Africa
  Rossing Uranium Ltd. (partially owned; see Rio Tinto Zinc)

The following companies are not listed in *Moody's Industrial Manual, 1976,* and therefore we have not been able to identify the corporations owning them or acting as Secretaries:
  Deelkraal Gold Mining Company, Ltd.
  Elandsrand Gold Mining Company, Ltd.
  Luipaards Vlei Estate and Gold Mining Company, Ltd.
  Randfontein Estates Gold Mining Company, Ltd.
  Southvaal Holdings Ltd.
  Vaal Reefs Exploration and Mining Company, Ltd.
  Western Deep Levels Gold Mining Company, Ltd.
  Western Reefs Exploration and Development Company, Ltd.

4. Publicly Owned Corporations and Commissions

Algeria

Societe Nationale de Recherche et d'Exploitation Miniere (SONAREM)

Australia

Australian Atomic Energy Commission (AAEC): the present policy of the AAEC does not favor direct investment in uranium mining. However, in the past the following holdings have existed:

A 6.5 percent share of the Ngalia joint venture, which is held by Urangesellschaft (q.v.), Ente Nazionale Idrocarburi, Magellan Petroleum Australia Ltd., Central Pacific Minerals N.L., and others
A 41.6 percent share of Mary Kathleen Uranium Ltd. (see Rio Tinto Zinc)
A 50 percent share of the Ranger joint venture, which is held by Peko-Wallsend, Ltd. (q.v.) and Electrolytic Zinc Corporation.

Brazil

Commissao Nacional de Energia Nuclear (CNEN)
Companhia Brasileira de Tecnologia Nuclear (CBTN)

Canada

Eldorado Nuclear Ltd.
Saskatchewan Mining and Development Corporation: joint venture in Saskatchewan with Inexco Oil Company and Uranerz Exploration and Mining Company, at Maurice Bay (50 percent SMDC, 25 percent Inexco, 25 percent Uranerz)
Societe Quebecois l'Exploration Miniere: joint venture in Quebec with Instituto Nacional de Industria (exploration phase) and Empresa Nacional del Uranio (development phase)
Uranium Canada Ltd.

France

Commissariat a l'Energie Atomique (CEA)
  Cogema
    Franco
  Uranex (one-third owned in 1975; established in 1969 by Compagnie de Mokta, Compagnie Francaise des Minerais d'Uranium, Pechiney Ugine Kuhlmann, and the CEA); joint ventures in Niger and Gabon with Minatome, which is a joint venture between Compagnie Francaise des Petroles and Pechiney Ugine Kuhlmann; joint venture in Saskatchewan, Canada, with Compagnie de Mokta, Compagnie Francaise des Minerais d'Uranium (q.v.), and Pechiney Ugine Kuhlmann; joint venture in Niger with Power Reactor and Nuclear Fuel Development Corporation (q.v.), the government of Niger, and Urangesellschaft; joint venture in Niger with Compagnie de Mokta, Compagnie Francaise des Minerais d'Uranium, Pechiney Ugine Kuhlmann, the government of Niger, Ente Nazionale Idrocarburi, and Urangesellschaft, called Societe des Mines de l'Air, or SOMAIR (33.5 percent CEA, 33.5 percent

Mokta-CFMU-Pechiney, 16.5 percent Niger, 8.125 percent ENI, 8.125 percent Urangesellschaft); joint venture in Niger with Overseas Uranium Resources Development Company and the government of Niger, called Compagnie Miniere d'Akouta (COMINAK); joint venture in Niger with Continental Oil Company

Italy

Ente Nazionale Idrocarburi (ENI)
  AGIP Mineraria
  AGIP Nucleare
    AGIP Nucleare Australia Pty Ltd.: joint venture in Australia with Urangesellschaft (q.v.), Magellan Petroleum Australia Ltd., Central Pacific Minerals N.L., and others
    Joint venture in Niger with the Commissariat a l'Energie Atomique (q.v.), Compagnie de Mokta, Compagnie Francaise des Minerais d'Uranium, Pechiney Ugine Kuhlmann, the government of Niger, and Urangesellschaft
    Somiren S.P.A.: joint venture in the Somali Republic with Overseas Uranium Resources Development Company

Niger

Uraniger and Onarem: joint venture in Niger with Power Reactor and Nuclear Fuel Development Corporation (q.v.), the Commissariat a l'Energie Atomique, and Urangesellschaft; joint venture in Niger with Commissariat a l'Energie Atomique (q.v.) Compagnie de Mokta, Compagnie Francaise des Minerais d'Uranium, Pechiney Ugine Kuhlmann, Ente Nazionale Idrocarburi, and Urangesellschaft; joint venture in Niger with the Commissariat a l'Energie Atomique (q.v.) and Overseas Uranium Resources Development Company; joint venture in Niger with Arabian Oil Company

Portugal

Junta de Energia Nuclear: joint ventures in Angola and Mozambique with Urangesellschaft

South Africa

Nuclear Fuels Corporation (NUFCOR)

Spain

Empresa Nacional del Uranio S.A. (ENUSA): joint venture in Quebec, Canada, with Societe Quebecois l'Exploration Miniere (development phase)
Instituto Nacional de Industria: joint venture in Quebec, Canada, with Societe Quebecois l'Exploration Miniere (exploration phase)

Sweden

A. B. Atomenergi
Statsforetag Aktiebolag
   Luossavaara Kiirunavaara A.B. (LKAB)

Bibliography

Legal Aspects of Uranium Development on Federal Lands

Davison, Robert P. "The Uranium Mining Lease." *Rocky Mountain Mineral Law Institute* 4 (1958):181–204.

Groves, James K. "Uranium Revisited." *Rocky Mountain Mineral Law Institute* 13 (1967):87–114.

Haggard, Jerry L. "Regulation of Mining Law Activities on Federal Lands." *Rocky Mountain Mineral Law Institute* 21 (1975):349–391.

Holbrook, R.B. "Legal Obstacles to Uranium Development." *Rocky Mountain Mineral Law Institute* 1 (1955):325–356.

Howard, Colin C. "Uranium Sales Contracts." *Rocky Mountain Mineral Law Institute* 22 (1969):389–403.

Lewis, Leonard J., and C. Keith Rooker. "Domestic Uranium Procurement—History and Problems." *Land and Water Law Review* 1,2 (1966):449–471.

McHenry, William B. "Uranium Concentrate Purchase Program, 1962–1966." *Atomic Energy Law Journal* 2 (Winter 1960):1–17.

Palmer, Robert S. "Problems Arising Out of Public Land Withdrawals of the Atomic Energy Commission." *Rocky Mountain Mineral Law Institute* 2 (1956):77–94.

Parcel, Randy L. "Federal, State and Local Regulation of Mining Exploration." *Rocky Mountain Mineral Law Institute* 22 (1976):405–432.

Parr, Clayton, J., and Ely Northcutt. "Mining Law." In *SME Mining Engineering Handbook*, ed. by Arthur B. Cummins and Ivan A. Given. New York: Society of Mining Engineers, 1973.

Peterson, Royal E. "The Uranium Royalty Provision: Its Evolution, Present Complexity and Future Uncertainty." *Rocky Mountain Mineral Law Institute* 22 (1976):865–889.

Root, Thomas E. "Legal Aspects of Mining by the In-Situ Leaching Method." *Rocky Mountain Mineral Law Institute* 22 (1976):349–387.

Strong, Kline D., and Warren O. Martin. "The Uranium Mining Lease." *Rocky Mountain Law Review* 27 (1955):425– . Reprinted in *Rocky Mountain Mineral Law Review* 4 (1967):107–132.

Swenson, Robert W. "Legal Aspects of Mineral Resource Exploitation." In *History of Public Land Law Development*, by Paul W. Gates and Robert W. Swenson. Washington: GPO, 1968.

U.S. Atomic Energy Commission. *Leasing of AEC Controlled Uranium Bearing Lands, Colorado, Utah, New Mexico*. Environmental Impact Statement, WASH-1523. Washington: USAEC, September 1972.

U.S. Congress. House. Committee on Interior and Insular Affairs. Subcommittee on Mines and Mining. *Multiple Use of Mineral Lands.* Hearings, 83d Cong., 2nd Sess., May 20, 21, 1954. Washington: GPO, 1954.

U.S. Congress. Senate. Committee on Interior and Insular Affairs. Subcommittee on Public Lands. *Multiple Mineral Use of Public Lands.* Hearings, 83d Cong., 2nd Sess., May 17–18, 1954. Washington: GPO, 1954.

U.S. Department of the Interior. Bureau of Indian Affairs. *Final Environmental Impact Statement: Navajo-Exxon Uranium Development.* FES 76-60. Billings, Mont.: USDI, 1976.

———. *Final Environmental Impact Statement: Sherwood Uranium Project, Spokane Indian Reservation.* FES 76/45. Portland, Ore.: USDI, August 1976.

———. *Final Environmental Statement of the Approval by the Department of the Interior of a Lease of the Ute Mountain Ute Tribal Lands for Uranium Exploration and Possible Mining.* FES-75-94. Albuquerque, New Mexico: USDI, November 21, 1975.

U.S. Federal Trade Commission staff. *Staff Report on Mineral Leasing on Indian Lands.* Washington: GPO, 1975.

———. *Report to the Federal Trade Commission on Federal Energy Land Policy: Efficiency, Revenue, and Competition.* Washington: GPO, 1976.

U.S. General Accounting Office. *Improvements Needed in Review of Public Land Withdrawals—Land Set Aside for Special Purposes.* CED-76-159. Washington: USGAO, November 16, 1976.

———. *Modernization of 1872 Mining Law Needed to Encourage Domestic Mineral Production, Protect the Environment, and Improve Public Land Management.* B-118678. Washington: USGAO, July 1974.

U.S. Nuclear Regulatory Commission, U.S. Department of the Interior, and U.S. Department of Agriculture. *Final Environmental Statement Related to Operation of Bear Creek Project, Rocky Mountain Energy Company.* NUREG-0129. Washington: USNRC, June, 1977.

Waldeck, William G. "Legal Problems Affecting Uranium Mining." In *Uranium and the Atomic Industry,* Proceedings of a Meeting of the Atomic Industrial Forum, Denver, Colorado, June 25 and 26, 1956. New York: AIF, 1956.

Watkins, T.H., and Charles S. Watson, Jr. *The Lands No One Knows: America and the Public Domain.* San Francisco: Sierra Club Books, 1975.

The Distribution of Uranium Lands and Federal Lands

U.S. Department of the Interior, Bureau of Land Management. *Public Land Statistics, 1975*. Washington: GPO, 1976.

U.S. Energy Research and Development Administeration. *Survey of Lands Held for Uranium Exploration, Development and Production in Fourteen Western States in the Six Month Period Ending June 30, 1975*. GJO-109 (76-1). Grand Junction, Colo.: USERDA, December 1975.

————. *Survey of Lands Held for Uranium Exploration, Development and Production in Fourteen Western States in the Six Month Period Ending June 30, 1976*. GJBX-2 (77). Grand Junction, Colo.: USERDA, 1977.

Domestic Uranium Demand

Baranowski, Frank P. "The Enriched Uranium Market." In *Nuclear Fuel Resources and Requirements*. WASH-1243. Washington: USAEC, April 1973.

Becker, Martin. "Testimony before the New York State Board on Electric Generation and the Environment," in the matter of Long Island Lighting Company—Jamesport Nuclear Units 1 and 2. 1977.

Berger, John J. *Nuclear Power: The Unviable Option*. Palo Alto, Calif.: Ramparts, 1976.

Bethe, Hans A. "Fuel for LWR, Breeders and Near-Breeders." *Annals of Nuclear Energy* 2 (1975):763–766.

Bethe, Hans A., and Chaim Braun. "Future Demand for Uranium Ore." In *Report of the Cornell Workshops on the Major Issues of a National Energy Research and Development Program*. Ithaca, N.Y.: College of Engineering, Cornell University, December 1973.

Bown, R.W., and R.H. Williamson. "Domestic Uranium Requirements." *Uranium Industry Seminar*, GJO-108(77). Grand Junction, Colo.: USDOE, 1977.

Day, M.C. "Nuclear Energy: A Second Round of Questions." *Bulletin of the Atomic Scientists* 31,10 (December 1975):52–59.

Hanrahan, Edward J. "Demand for Uranium and Separative Work." Paper presented at AIF Fuel Cycle Conference '76, Phoenix, Arizona, March 22, 1976.

————. "Domestic and Foreign Uranium Requirements, 1975–2000." In

Uranium Industry Seminar. GJO-108(75). Grand Junction, Colo.: USERDA, 1975.

————. "Nuclear Power Growth, Enrichment and Uranium Requirements, 1974 to Year 2000." In *Uranium Industry Seminar,* GJO-108(74). Grand Junction, Colo.: USAEC, 1974.

Hanrahan, Edward J., Richard H. Williamson, and Robert W. Bown. "The Changing Nuclear Picture: Uranium and Separative Work Requirements." Paper presented at Atomic Industrial Forum Fuel Cycle Conference '77, Kansas City, Mo., April 25, 1977.

————. "United States Uranium Requirements." In *Uranium Industry Seminar,* GJO-108(76). Grand Junction, Colo.: USERDA, 1976.

Hogerton, John F. "U.S. Uranium Requirements." In *Uranium Supply and Demand,* ed. by M.J. Spriggs and K.D. Casteel. London: Mining Journal Books, 1977.

Houston, Dean. "Testimony Regarding Light Water Reactor Fuel Performance Prepared by D. Houston." Before the Atomic Safety and Licensing Board, in the matter of Pacific Gas and Electric Co. (Diablo Canyon 1 and 2), 1976.

Kazmann, Raphael G. "Do We Have a Nuclear Option?" *Mining Engineering* 27,8 (August 1975):35–37.

Luce, C.F. "Whither North American Demand for Uranium?" In *Uranium Supply and Demand,* ed. by Michael J. Spriggs. London: Uranium Institute, 1976.

Parks, Joe W., and David C. Thomas. "Plans for Operating Enrichment Plants and the Effect on Uranium Supply." In *Uranium Industry Seminar,* GJO-108(76). Grand Junction, Colo.: USERDA, 1976.

Schwennesen, Jarvis L. "Overview of the Nuclear Fuel Cycle." In *Uranium Industry Seminar,* GJO-108(76). Grand Junction, Colo.: USERDA, 1976.

Sondermayer, Roman V. "Thorium." In U.S. Department of the Interior, Bureau of Mines, *Minerals Yearbook, 1973.* Wshington: GPO, 1975.

Thomas, D.C. "Future Relationship of Uranium Supply and Enrichment." In *Uranium Industry Seminar,* GJO-108(75). Grand Junction, Colo.: USERDA, 1975.

Thornton, C.D.W. "The Uranium Producer Plans for Atomic Power." In *Uranium and the Atomic Industry,* proceedings of a meeting of the Atomic Industrial Forum, Denver, Colorado, June 25 and 26, 1956. New York: AIF, 1956.

U.S. Energy Research and Development Administration. *Nuclear Fuel Cycle.* A Report by the Fuel Cycle Task Force. Washington: USERDA, March 1975.

U.S. General Accounting Office. *Allocation of Uranium Enrichment Services to Fuel Foreign and Domestic Nuclear Reactors.* Report to the

Committee on Foreign Affairs, House of Representatives, ID-75-45. Washington: USGAO, March 4, 1975.

————. *Comments on Proposed Legislation to Change Basis for Government Charge for Uranium Enrichment Services,* RED-76-30. Washington: USGAO, September 22, 1977.

U.S. Nuclear Regulatory Commission. *Final Generic Environmental Statement on the Use of Recycle Plutonium in Mixed Oxide Fuel in Light Water Cooled Reactors,* NUREG-0002. Washington: USNRC, August 1976.

Vernon, John M. *Public Investment Planning in Civilian Nuclear Power.* Durham, N.C.: Duke Univ. Press, 1971.

Wood, P.M. "NRC Staff Testimony on Uranium Fuel Use Efficiency," in the matter of Gulf States Utilities Company (River Bend Station, Units 1 and 2). Washington: USNRC, 1976.

Projections of Installed Nuclear Capacity in the United States

U.S. Atomic Energy Commission. "Probable Course of Industrial Development of Economic Nuclear Power." Statement by AEC before the Joint Committee on Atomic Energy at Hearings on Proposed Amendments to the Atomic Energy Act, June 2, 1954.

Cohen, Karl. "Charting a Course for Nuclear Power Development." *Nucleonics* 16,1 (January 1958):66–70.

Warren, Frederick H., William W. Lowe, and James K. Pickard. *A Growth Survey of the Atomic Industry, 1958–1968.* New York: Atomic Industrial Forum, February 1958.

U.S. Atomic Energy Commission. *Civilian Nuclear Power—A Report to the President.* Washington: USAEC, December 1962.

————. *Estimated Growth of Civilian Nuclear Power,* WASH-1055. Washington: GPO, 1965.

————. Press Release S-20-66. Washington: USAEC, June 7, 1966.

————. Press Release S-23-66. Washington: USAEC, September 8, 1966.

————. *Forecast of Growth of Nuclear Power,* WASH-1084. Washington: USAEC, December 1967.

————. *Civilian Nuclear Power—The 1967 Supplement to the 1962 Report to the President.* Washington: USAEC, 1967.

United Engineers and Constructors Inc. and S.M. Stoller Associates. *Current Status and Future Technical and Economic Potential of Light Water Reactors,* WASH-1082. Washington: USAEC, March 1968.

U.S. Atomic Energy Commission. Forecast of Nuclear Power Growth. Unpublished, May 1969. Described on page 8 of WASH-1139.

————. *Potential Nuclear Power Growth Patterns,* WASH-1098. Washington: USAEC, December 1970.

————. *Forecast of Growth of Nuclear Power,* WASH-1139. Washington: USAEC, January 1971.

Faulkner, Rafford L. "Outlook for Uranium Production to Meet Future Nuclear Fuel Needs in the United States." Paper presented at the Fourth United Nations International Conference on the Peaceful Uses of Atomic Energy, Geneva, Switzerland, 6–16 September 1971. Dated May 1971.

U.S. Atomic Energy Commission. *The Growth of Nuclear Power, 1972– 1985,* WASH-1139 (Rev. 1). Washington: USAEC, December 1971.

————. *Nuclear Power, 1973–2000,* WASH-1139(72). Washington: USAEC, December 1972.

————. *Nuclear Power Growth 1974–2000,* WASH-1139(74). Washington: USAEC, February 1974.

U.S. Energy Research and Development Administration. Update of WASH-1139(74). Unpublished, February 1975. Described on page III(B)-2 of U.S. Nuclear Regulatory Commission, *Final Generic Environmental Statement on the Use of Recycle Plutonium in Mixed Oxide Fuel in Light Water Cooled Reactors,* NUREG-0002. Washington: USNRC, August 1976.

Hanrahan, Edward J. "Domestic and Foreign Uranium Requirements, 1975–2000." In *Uranium Industry Seminar,* GJO-108(75). Grand Junction, Colo.: USERDA, 1975.

————. "Demand for Uranium and Separative Work." Paper presented at AIF Fuel Cycle Conference '76, Phoenix, Arizona, March 22, 1976.

Joskow, P.L., and M.L. Baughman. "The Future of the U.S. Nuclear Energy Industry." *Bell Journal of Economics* 7,1 (Spring 1976):3–32.

Davis, W. Kenneth. Letter to the Editor. *Scientific American* 234,4 (April 1976):8.

Hanrahan, Edward J., Richard H. Williamson, and Robert W. Bown. "United States Uranium Requirements." In *Uranium Industry Seminar,* GJO-108(76). Grand Junction, Colo.: USERDA, 1976.

————. "The Changing Nuclear Picture: Uranium and Separative Work Requirements." Paper presented at Atomic Industrial Forum Fuel Cycle Conference '77, Kansas City, Missouri, April 25, 1977.

Bown, R.W., and R.H. Williamson. "Domestic Uranium Requirements." *Uranium Industry Seminar,* GJO-108(77). Grand Junction, Colo.: USDOE, 1977.

Uranium Production Technology

Arthur D. Little, Inc. *An Assessment of the Economic Effects of Radiation Exposure Standards for Uranium Miners*. Cambridge, Mass.: ADL, September 1970.

Battelle–Pacific Northwest Laboratories. *Assessment of Uranium and Thorium Resources in the United States and the Effect of Policy Alternatives*. Prepared for National Science Foundation. Springfield, Va.: NTIS, December 1974.

Bennett, Harold J., Jerrold G. Thompson, Herbert J. Quiring, and Joseph E. Toland. *Financial Evaluation of Mineral Deposits Using Sensitivity and Probabilistic Analysis Methods*. Information Circular 8495, U.S. Bureau of Mines. Washington: GPO, 1970.

Biles, Martin B., and Richard H. Kennedy. "Land Use Problems of Uranium Mill Tailings." Paper presented at AIF Conference on Land Use and Nuclear Facility Siting, Denver, Colorado, July 18–21, 1976.

Clark, D.A. *State of the Art: Uranium Mining, Milling and Refining Industry*, EPA-660/2-74-038. Corvallis, Ore.: U.S. Environmental Protection Agency, National Environmental Research Center, June 1974.

Clegg, John W., and Dennis D. Foley. *Uranium Ore Processing*. Prepared under contract with USAEC. Reading, Mass.: Addison-Wesley, 1958.

Colorado Department of Health. *Uranium Wastes and Colorado's Environment*. Denver, Colo.: CDH, August 1971.

Comey, David D. "The Legacy of Uranium Tailings." *Bulletin of the Atomic Scientists* 31,7 (September 1975):43–45.

Cook, Lewis M., Bradley W. Caskey, and Martin C. Wukasch. "The Effects of Uranium Mining on Environmental Gamma Ray Exposures." *Proceedings, IVth International Radiation Protection Association*. Paris, 1977.

Cummins, Arthur B., and Ivan A. Given. *SME Mining Engineering Handbook*. New York: Society of Mining Engineers, 1973.

Duncan, D.L., and G.G. Eadie. *Environmental Surveys of the Uranium Mill Tailings Piles and Surrounding Areas, Salt Lake City, Utah*, Report EPA-520/6-74-006. Washington: USEPA, August 1974.

Frank, Jacob N. "Cost Model for Solution Mining of Uranium." In USERDA, *Uranium Industry Seminar*, GJO-108(76). Grand Junction, Colo.: USERDA, 1976.

Goldsmith, W.A. "Radiological Aspects of Inactive Uranium-Milling Sites: An Overview." *Nuclear Safety* 17,6 (November-December 1976):722–732.

Goodwin, Aurel. "Problems and Techniques for Removal of Radon and Radon-Daughter Products from Mine Atmospheres." *Nuclear Safety* 14,6 (November-December 1973):643–653.

Grimes, W.R. "Natural Resource Use In and Effluents from Mining and Milling Operations for Various Nuclear Reactors." Oak Ridge, Tenn.: Oak Ridge National Laboratory, October 28, 1976.

Harris, William B. "Permissible Limits for Radon Concentrations in Uranium Mining." In *Uranium and the Atomic Industry,* proceedings of a meeting of the Atomic Industrial Forum, Denver, Colorado, June 25, and 26, 1956. New York: AIF, 1956.

Haskell, Floyd K. "Federal Income Taxation of the Uranium Industry." *Rocky Mountain Law Review* 27 (1955):469–481.

Hollocher, Thomas, and James MacKenzie. "Radiation Hazards Associated with Uranium Mill Operations." In D.F. Ford et al., *The Nuclear Fuel Cycle.* Cambridge, Mass.: M.I.T. Press, 1975.

International Atomic Energy Agency. *Management of Wastes from the Mining and Milling of Uranium and Thorium Ores.* Safety Series No. 44. Vienna: IAEA, 1976.

Klemenic, John A. "An Estimate of the Economics of Uranium Concentrate Production from Low Grade Sources." Grand Junction, Colo.: USAEC, 1974.

———. "Examples of Overall Economics in a Future Cycle of Uranium Concentrate Production for Assumed Open Pit and Underground Mining Operations." Grand Junction, Colo.: USAEC, 1972.

Lowell, A.F. "Factors Affecting Mining Costs." In *Uranium Supply and Demand,* ed. by M.J. Spriggs and K.D. Casteel. London: Mining Journal Books, 1977.

Lyon, Barbara. "Uranium Tailings in the Public Eye." *Oak Ridge National Laboratory Review* 3,6 (Summer 1976):18–23.

Martin, John B. "NRC Uranium Mill Licensing Activities." In USERDA, *Uranium Industry Seminar,* GJO-108(76). Grand Junction, Colo.: USERDA, 1976.

Maxfield, Peter C. *The Income Taxation of Mining Operations.* Boulder, Colo.: Rocky Mountain Mineral Law Foundation, 1975.

Metzger, H. Peter. *The Atomic Establishment.* New York: Simon and Schuster, 1972.

———. " 'Dear Sir: Your House Is Built On Radioactive Uranium Waste.' " *New York Times Magazine,* October 31, 1971.

Miller, Ralph L., and James R. Gill. "Uranium From Coal." *Scientific American* 191,4 (October 1954):36–39.

Pohl, Robert O. "Coal Ash—A New Source of Uranium?" Ithaca, N.Y., 1976.

————. "Health Effects of Radon-222 from Uranium Mining." *Search* 7,8 (August 1976):345–354. With comments by R.M. Fry and J.E. Cook.

————. "Land Use for Nuclear and for Solar Energy." Ithaca, N.Y., October 1976.

Ramsay, William. "Radon from Uranium Mill Tailings: A Source of Significant Radiation Hazard?" *Environmental Management* 1,2 (1976):139–145.

Ripley, Anthony. "The AEC in Colorado." *The Washington Monthly* 2,5 (July 1970):7–14.

Rowe, W.D., F.L. Galpin, and H.T. Peterson, Jr. "EPA's Environmental Radiation-Assessment Program." *Nuclear Safety* 16-6 (November-December 1975):667–682.

Schurgin, A., and Thomas Hollocher. "Radiation-Induced Lung Cancers Among Uranium Miners." In D.F. Ford et al., *The Nuclear Fuel Cycle*. Cambridge, Mass.: M.I.T. Press, 1975.

Sears, M.B., R.E. Blanco, R.C. Dahlman, G.S. Hill, A.D. Ryon, and J.P. Witherspoon. *Correlation of Radioactive Waste Treatment Costs and the Environmental Impact of Waste Effluents in the Nuclear Fuel Cycle for Use in Establishing 'As Low as Practicable' Guides— Milling of Uranium Ores*, ORNL-TM-4903. Oak Ridge, Tenn.: Oak Ridge National Laboratory, May 1975.

Swift, J.J., J.M. Hardin, and H.W. Calley. *Potential Radiological Impact of Airborne Releases and Direct Gamma Radiation to Individuals Living Near Inactive Uranium Mill Tailings Piles*. Report EPA-520/1-76-001. Washington: USEPA, January 1976.

U.S. Atomic Energy Commission. *Draft Detailed Statement on the Environmental Considerations by the U.S. Atomic Energy Commission, Fuels and Materials, Directorate of Licensing, Related to the Proposed Issuance of a License to the Rio Algom Corporation for the Humeca Uranium Mill*, Docket No. 40-8084. Washington: USAEC, December 1972.

————. *Final Environmental Statement Related to Operation of the Highland Uranium Mill by the Exxon Company, U.S.A.*, Docket No. 40-8102. Washington: USAEC, March 1973.

————. *Final Environmental Statement Related to Operation of Shirley Basin Uranium Mill, Utah International, Inc.*, Docket No. 40-6622. Washington: USAEC December 1974.

U.S. Congress. House. Committee on Education and Labor. Selection Subcommittee on Labor. *Uranium Miners Compensation*. Hearings, 90th Cong., 2d Sess., April 1, 4, and May 1, 1968. Washington: GPO, 1968.

U.S. Congress. Joint Committee on Atomic Energy. Subcommittee on

Research, Development, and Radiation. *Radiation Exposure of Uranium Miners*. Summary analysis of hearings, May 9, 10, 23, June 6, 7, 8, 9, July 26, 27 and August 8, 10, 1967. Washington: GPO, 1968.

———. Subcommittee on Research, Development, and Radiation. *Radiation Standards for Uranium Mining*. Hearings, 91st Cong., 1st Sess., March 17, 18, 1969. Washington: GPO, 1969.

———. Subcommittee on Raw Materials. *S.2566 and H.R. 11378: Uranium Mill Tailings in the State of Utah*. Hearings, 93rd Cong., 2d Sess., March 12, 1974. Appendix 3 to hearings on ERDA authorizing legislation for FY1976. Washington: GPO, 1975.

———. *Use of Uranium Mill Tailings for Construction Purposes*. Hearing, 92d Cong., 1st Sess., October 28 and 29, 1971. Washington: GPO, 1971.

U.S. Congress. Senate. Committee on Public Works. Subcommittee on Air and Water Pollution. *Radioactive Water Pollution in the Colorado River Basin*. Hearing, 89th Cong., 2d Sess., May 6, 1966. Washington: GPO, 1966.

U.S. Energy Research and Development Administration. *Conference on Occupational Health Experience With Uranium*, ERDA 93. Washington: GPO, 1976.

U.S. Environmental Protection Agency. *Environmental Analysis of the Uranium Fuel Cycle*, Report EPA-520/9-73-3. Washington: USEPA, 1973.

———. *Radium-226, Uranium and Other Radiological Data from Water Quantity Surveillance Stations Located in the Colorado River Basin of Colorado, Utah, New Mexico and Arizona, January 1961 through June 1972*. Denver, Colo.: USEPA, July 1973.

———. *Water Quality Impacts of Uranium Milling and Mining Activities in the Grants Mineral Belt, New Mexico*. Dallas, Tex.: USEPA, September 1975.

U.S. Environmental Protection Agency and U.S. Atomic Energy Commission. *Studies of Inactive Uranium Mill Sites and Tailings Piles*. Summary Report, Phase I. Washington: USEPA, October 1974.

U.S. General Accounting Office. *Controlling the Radiation Hazard from Uranium Mill Tailings*, RED-75-365. Washington: USGAO, May 21, 1975.

———. *After Years of Effort, Accident Rates Are Still Unacceptably High in Mines Covered by the Federal Metal and Nonmetallic Mine Safety Act*, CED-77-103. Washington: USGAO, July 26, 1977.

U.S. Nuclear Regulatory Commission. *Final Environmental Statement Related to Operation of the Humeca Uranium Mill by the Rio Algom Corporation*, NUREG-0046. Washington: USNRC, April 1976.

———. *Draft Environmental Statement Related to Operation of Lucky*

Mc Uranium Mill, Utah International, Inc., NUREG-0295. Washington: USNRC, June 1977.

——. *Final Environmental Statement Related to the Operation of the Humeca Uranium Mill*, NUREG-0046. Washington: USNRC, April 1976.

Youngberg, Elton A. "The Uranium Industry Industry: Exploration, Mining, and Milling." *IEEE Transactions on Power Apparatus and Systems.* (July–August 1973):1201–1208.

The Domestic Uranium Resource Base

Adams, S.S. "Problems in the Conversion of Uranium Resources to Uranium Reserves." In *Mineral Resources and the Environment. Supplementary Report: Reserves and Resources of Uranium in the United States,* by NAS-NRC Committee on Mineral Resources and the Environment. Washington: National Academy of Sciences, 1975.

Adler, H.H. "Geological Aspects of Foreign and Domestic Deposits and Their Bearing on Exploration." In *Uranium Industry Seminar,* GJO-108(75). Grand Junction, Colo.: USERDA, 1975.

Armstrong, Frank C. "Alternatives to Sandstone Deposits." In *Mineral Resources and the Environment. Supplementary Report: Reserves and Resources of Uranium in the United States,* by NAS-NRC Committee on Mineral Resources and the Environment. Washington: National Academy of Sciences, 1975.

Bieniewski, C.L., F.H. Persse, and E.F. Brauch. *Availability of Uranium at Various Prices from Resources in the United States.* U.S. Bureau of Mines Information Circular 8501, 1971.

Blanc, Robert P. "U.S. Uranium Potential Resources—Fact or Fiction." In *Mineral Resources and the Environment. Supplementary Report: Reserves and Resources of Uranium in the United States,* by NAS-NRC Committee on Mineral Resources and the Environment. Washington: National Academy of Sciences, 1975.

Bryant, E.A., G.A. Cowan, W.R. Daniels, and W.J. Maeck. "Oklo, an Experiment in Long-Term Geologic Storage." ACS Symposium Series, No. 35. *Actinides in the Environment.* (American Chemical Society, 1976):89–102.

Bowyer, Ben. "Quantity and Quality of Data Available on Potential Resources of Uranium." In *Mineral Resources and the Environment. Supplementary Report: Reserves and Resources of Uranium in the United States,* by NAS-NRC Committee on Mineral Resources and the Environment. Washington: National Academy of Sciences, 1975.

Butler, Arthur P. "Geological Basis for Estimation of Potential Resources in Sandstone Deposits." In *Mineral Resources and the Environment. Supplementary Report: Reserves and Resources of Uranium in the United States*, by NAS-NRC Committee on Mineral Resources and the Environment. Washington: National Academy of Sciences, 1975.

Cochran, Thomas B. *The Liquid Metal Fast Breeder Reactor: An Environmental and Economic Critique*. Washington: Resources for the Future, Inc., 1974.

Cowan, George A. "A Natural Fission Reactor." *Scientific American* (July 1976):36–47.

Darnley, A.G. "Geophysics in Uranium Exploration." In *Uranium Exploration '75*, by Geological Survey of Canada. Ottawa: Supply and Services Canada, 1976.

Davis, James F. "Problems Involved in Converting Resources to Reserves." In *Mineral Resources and the Environment. Supplementary Report: Reserves and Resources of Uranium in the United States*, by NAS-NRC Committee on Mineral Resources and the Environment. Washington: National Academy of Sciences, 1975.

Douglas, Richard F. "Government-Industry Collaboration and Related Problems in the Preparation of Uranium Resource Estimates." In *Mineral Resources and the Environment. Supplementary Report: Reserves and Resource of Uranium in the United States*, by NAS-NRC Committee on Mineral Resources and the Environment. Washington: National Academy of Sciences, 1975.

Dyck, Willy. "Geochemistry Applied to Uranium Exploration." In *Uranium Exploration '75*, by Geological Survey of Canada. Ottawa: Supply and Services Canada, 1976.

Electric Power Research Institute. *Uranium Resources to Meet Long Term Uranium Requirements*, EPRI SR-5. Palo Alto, Calif.: EPRI, November 1974.

Ellis, J.R., D.P. Harris, and N.H. Vanwie. *A Subjective Probability Appraisal of Uranium Resources in New Mexico*, GJO-110(76). Grand Junction, Colo.: USERDA, 1976.

Finch, W.I., A.P. Butler, Jr., and F.C. Armstrong. "Nuclear Fuels." In *United States Mineral Resources*, ed. by D.A. Probst and W.A. Pratt. U.S. Geological Survey Professional Paper 820. Reston, Va.: USGS, 1973.

Grutt, Eugene W., Jr. "Cooperation Between the AEC and the Uranium Industry." In *Mineral Resources and the Environment. Supplementary Report: Reserves and Resources of Uranium in the United States*, by NAS-NRC Committee on Mineral Resources and the Environment. Washington: National Academy of Sciences, 1975.

Harris, DeVerle P. *The Estimation of Uranium Resources by Life-Cycle or Discovery-Rate Models—A Critique,* GJO-112(76). Grand Junction, Colo.: USERDA, 1976.

————. "Geostatistics in the Appraisal of Metal Resources." In *Mineral Materials Modeling: A State-of-the-Art Review,* ed. by William A. Vogely. Washington: Resources for the Future, December 1975.

————. *A Survey and Critique of Quantitative Methods for the Appraisal of Mineral Resources,* GJBX-12(76). Grand Junction, Colo.: USERDA, 1976.

Hetland, Donald L. "Discussion of the Preliminary NURE Report and Potential Resources." In *Uranium Industry Seminar,* GJO-108(76). Grand Junction, Colo.: USERDA, 1976.

Hetland, Donald L. "Potential Resources." In *Uranium Industry Seminar,* GJO-108(75). Grand Junction, Colo.: USERDA, 1975.

Hogerton, J.F., C.H. Barnes, L. Geller, and D.R. Hill. *Uranium Data,* EPRI EA-400. Palo Alto, Calif.: Electric Power Research Institute, 1977.

Holdren, John P. "Uranium Availability and the Breeder Decision." *Energy Systems and Policy* 1,3 (1975):205–232.

International Atomic Energy Agency. *Uranium Exploration Geology.* Proceedings of a Panel Held in Vienna, April 13–17, 1970. Vienna: IAEA, 1970.

Klemenic, John, and Laurence Sanders. "Discovery Rates and Costs for Additions to Constant-Dollar Uranium Reserves." In *Uranium Industry Seminar,* GJO-108(76). Grand Junction, Colo.: USERDA, 1976.

Lieberman, M.A. "United States Uranium Resources—An Analysis of Historical Data." *Science* 192 (April 30, 1976):431–436.

Malan, Roger C. "NURE-Framework, Scope and Progress." In *Uranium Industry Seminar,* GJO-108(75). Grand Junction, Colo.: USERDA, 1975.

————. "Status of the NURE Program." In *Uranium Industry Seminar,* GJO-108(76). Grand Junction, Colo.: USERDA, 1976.

Motica, John E. "Problems Involved in Converting Uranium Resources to Reserves." In *Mineral Resources and the Environment. Supplementary Report: Reserves and Resources of Uranium in the United States,* by NAS-NRC Committee on Mineral Resources and the Environment. Washington: National Academy of Sciences, 1975.

National Academy of Sciences. National Research Council. Commission on Natural Resources. Committee on Mineral Resources and the Environment (COMRATE). *Mineral Resources and the Environment. Supplementary Report: Reserves and Resources of Uranium in the United States.* Washington: NAS, 1975.

Nininger, Robert D. *Exploration for Nuclear Raw Materials*. New York: D. Van Nostrand, 1956.

―――. "Uranium Resources." Paper presented at American Institute of Chemical Engineers, Annual Meeting, Salt Lake City, Utah, August 19–21, 1974.

――― and S.H.U. Bowie. "Technological Status of Nuclear Fuel Resources." Paper presented at ANS/ENS Conference on World Nuclear Energy, Washington, November 15, 1976.

Nuclear Energy Policy Study Group. *Nuclear Power Issues and Choices*. Cambridge, Mass.: Ballinger, 1977.

Rackley, Ruffin I. "Problems of Converting Potential Uranium Resources into Mineable Reserves." In *Mineral Resources and the Environment. Supplementary Report: Reserves and Resources of Uranium in the United States,* by NAS-NRC Committee on Mineral Resources and the Environment. Washington: National Academy of Sciences, 1975.

Roach, C.H. "Contracted Investigations and Information Outputs." In *Uranium Industry Seminar,* GJO-108(75). Grand Junction, Colo.: USERDA, 1975.

Roach, Carl H. "Uranium Geophysical Technology Development." In *Uranium Industry Seminar,* GJO-108(76). Grand Junction, Colo.: USERDA, 1976.

Searl, Milton F., and Jeremy Platt. "Views on Uranium and Thorium Resources." *Annals of Nuclear Energy* 2 (1975):751–762.

Singer, D.A. "Mineral Resource Models and the Alaskan Mineral Resource Assessment Program." In *Mineral Materials Modeling: A State-of-the-Art Review*. Washington: Resources for the Future, 1975.

Southern Interstate Nuclear Board. *Uranium in the Southern United States,* WASH-1128. Washington: U.S. Atomic Energy Commission, July 1969.

Torries, Thomas F. "A Probabilistic Model of the Availability of Domestically Produced U_3O_8 in the United States." In *Mineral Materials Modeling: A State-of-the-Art Review,* ed. by William A. Vogely. Washington: Resources for the Future, 1975.

U.S. Atomic Energy Commission. Grand Junction Office. *Nuclear Fuel Resource Evaluation: Concepts, Uses, Limitations,* GJO-105. Grand Junction, Colo.: USAEC, May 1973.

U.S. Department of the Interior. Geological Survey. "An Assessment of Potential U.S. Uranium Supply." Washington: USDI, 1974.

――― et al. *Mineral and Water Resources of Alaska*. Report to the Senate Committee on Interior and Insular Affairs. Washington: GPO, 1964.

————. *Mineral and Water Resources of New Mexico.* Report to the Senate Committee on Interior and Insular Affairs. Washington: GPO, 1965.

————. *Mineral and Water Resources of Arizona.* Report to the Senate Committee on Interior and Insular Affairs. Washington: GPO, 1969.

————. *Mineral and Water Resources of Colorado.* Report to the Senate Committee on Interior and Insular Affairs. Washington: GPO, 1964.

U.S. Energy Research and Development Administration. *National Uranium Resource Evaluation, Preliminary Report,* GJO-111 (76). Grand Junction, Colo.: USERDA, June 1976.

————. *Survey of Uranium Industry Views Concerning U.S. Uranium Resources and U.S. Uranium Production Activities,* ERDA 77-59. Springfield, Va.: NTIS, 1977.

U.S. General Accounting Office. *Domestic Energy Resource and Reserve Estimates—Uses, Limitations, and Needed Data,* EMD-77-6. Washington: USGAO, March 17, 1977.

Yarger, Larry L. "Potential Resources—A Utility Viewpoint." In *Mineral Resources and the Environment. Supplementary Report: Reserves and Resources of Uranium in the United States,* by NAS-NRC Committee on Mineral Resources and the Environment. Washington: National Academy of Sciences, 1975.

Zitting, R.T. "Estimation of Potential Uranium Resources." In *Mineral Resources and the Environment. Supplementary Report: Reserves and Resources of Uranium in the United States,* by NAS-NRC Committee on Mineral Resources and the Environment. Washington: National Academy of Sciences, 1975.

The Domestic Market

Atomic Industrial Forum, Inc. *The Domestic Uranium Raw Materials Industry.* A Status Report by the Committee on Mining and Milling. New York: AIF, May 1962.

————. *Uranium and the Atomic Industry.* Proceedings of a Meeting of the Atomic Industrial Forum, Denver, Colorado, June 25 and 26, 1956. New York: AIF, 1956.

Baroch, Charles T., and Charles J. Baroch. "Nuclear Energy Minerals and Their Utilization." In *Economics of the Mineral Industries,* ed. by William A. Vogely. New York: American Institute of Mining, Metallurgical, and Petroleum Engineers, 1976.

Boyd, James. "United States Uranium Position." In *Uranium Supply and Demand,* ed. by M.J. Spriggs and K.D. Casteel. London: Mining Journal Books, 1977.

Chenoweth, William L. "Exploration Activities." In *Uranium Industry Seminar,* GJO-108(76). Grand Junction, Colo.: USERDA, 1976.

Cherry, B.H. "Uranium—The Market and Adequacy of Supply." In *Mineral Resources and the Environment. Supplementary Report: Reserves and Resources of Uranium in the United States,* by NAS-NRC Committee on Mineral Resources and the Environment. Washington: National Academy of Sciences, 1975.

Cobleigh, Ira U. *A Killing in Uranium.* Long Island City, N.Y.: DuVal's Consensus, 1954.

Facer, J. Fred, Jr. "Production Statistics." In *Uranium Industry Seminar,* GJO-108(76). Grand Junction, Colo.: USERDA, 1976.

————. "Production Statistics." In *Uranium Industry Seminar,* GJO-108(75). Grand Junction, Colo.: USERDA, 1975.

Faulkner, Rafford L. "Outlook for Uranium Production to Meet Future Nuclear Fuel Needs in the United States." Paper presented at Fourth U.N. International Conference on the Peaceful Uses of Atomic Energy, Geneva, September 6–16, 1971.

————. "Uranium Supply and Demand." *Atomic Energy Law Journal* 11 (1969):131–143.

Federal Energy Resources Council. "Uranium Reserves, Resources, and Production." Washington: FERC, June 15, 1976.

Geller, Leonard. "Testimony before the New York Public Service Commission in Case No. 26974," Generic Proceeding on Nuclear/Fossil Generation, November 29, 1976.

Hogerton, John F. "U.S. Uranium Supply Outlook." In *Uranium Supply and Demand,* ed. by M.J. Spriggs and K.D. Casteel. London: Mining Journal Books, 1977.

————, Clifton H. Barnes, William A. Franks, and Robert J. McWhorter. *Report on Uranium Supply: Task III of EEI Nuclear Fuels Supply Study Program.* December 5, 1975. New York: S.M. Stoller, 1975.

Hogerton, John F., Robert J. McWhorter, and Joel E. Gingold. *Assessment of the United States Uranium Supply Outlook, Short and Medium Term (1977–1996).* New York: S.M. Stoller, 1977.

Johnson, Jesse C. "The AEC's Program and Policies in the Uranium Field." In *Uranium and the Atomic Industry,* Proceedings of a Meeting of the Atomic Industrial Forum, Denver, Colorado, June 25 and 26, 1956. New York: AIF, 1956.

Johnson, Wilfrid E. "AEC Uranium Policies." *Atomic Energy Law Journal* 12 (1970):263–273.

Joskow, Paul L. "Commercial Impossibility, the Uranium Market and the Westinghouse Case." *Journal of Legal Studies* 6,1 (January 1977): 119–176.

Klemenic, John. "Analysis and Trends in Uranium Supply." In *Uranium*

Industry Seminar, GJO-108(76). Grand Junction, Colo.: USERDA, 1976.

———. "Uranium Supply and Associated Economics: A Fifteen-Year Outlook." In *Uranium Industry Seminar,* GJO-108(75). Grand Junction, Colo.: USERDA, 1975.

Lapp, Ralph E. "We May Find Ourselves Short of Uranium, Too." *Fortune* (October 1975):151–199.

McWhorter, R.J., D.R. Hill, E.A. Noble, J.F. Hogerton, and C.H. Barnes. *Uranium Exploration Activities in the United States,* EPRI EA-401. Palo Alto, Calif.: Electric Power Research Institute, 1977.

Meehan, Robert J. "Uranium Reserves and Exploration Activity." In *Uranium Industry Seminar,* GJO-108(75). Grand Junction, Colo.: USERDA, 1975.

———.. "Uranium Reserves and Exploration." In *Uranium Industry Seminar,* GJO-108(76). Grand Junction, Colo.: USERDA, 1976.

Mulholland, Joseph P., and Douglas W. Webbink. *Concentration Levels and Trends in the Energy Sector of the U.S. Economy.* Staff Report to the Federal Trade Commission. Washington: GPO, 1974.

National Petroleum Council. *U.S. Energy Outlook: Nuclear Energy Availability.* Washington: National Petroleum Council, 1973.

Nelson, William H. "Mining and Marketing the Uranium." *Rocky Mountain Law Review* 27 (1955):482–495.

Nininger, Robert D. "Uranium Reserves and Requirements." In *Nuclear Fuel Resources and Requirements,* WASH-1243. Washington: USAEC, April 1973.

———. "Current Uranium Supply Picture." Paper presented at the Conference on Nuclear Power and Applications in Latin America, Mexico City, September 28–October 1, 1975.

Nuclear Exchange Corporation. "Significant Events in the Uranium Market: 1969–1976." *Nuclear News* 19,15 (December 1976):46–51.

Patterson, John A. "Outlook for Uranium." Paper presented at the 17th Minerals Symposium, AIMMPE, Casper, Wyoming, May 11, 1974.

———. "Status of Uranium Procurement." Paper presented at AIF Fuel Cycle Conference '77, Kansas City, Missouri, April 25, 1977.

———. "Uranium Market Activities." In *Uranium Industry Seminar,* GJO-108(76). Grand Junction, Colo.: USERDA, 1976.

———. "Uranium Market Activities." In *Uranium Industry Seminar,* GJO-108(75). Grand Junction, Colo.: USERDA, 1975.

———. "Uranium Market Activities." In *Nuclear Fuel Resources and Requirements,* WASH-1243. Washington: USAEC, April 1973.

———. "Uranium Supply Development." Paper presented at AIF Fuel Cycle Conference '76, Phoenix, Arizona, March 22, 1976.

———. "U.S. Uranium Supply and Demand Overview." Paper presented at American Nuclear Society Executive Conference on Uranium Supply, Monterey, California, January 24, 1977.

――――. "U.S. Uranium Supply Position." Statement Before Connecticut Public Utility Control Authority, Hartford, Connecticut, March 9, 1976.

――――. "Uranium Requirements and Supply Outlook." Washington: USERDA, November 1975.

Public Finance Research Program, University of New Mexico. *Uranium Industry in New Mexico,* Report No. 76-100B. Albuquerque, N.M.: Univ. of New Mexico Energy Resources Board, 1976.

Sanger, Herbert S., Jr., and William E. Mason. *The Structure of the Energy Markets: A Report of TVA's Antitrust Investigation of the Coal and Uranium Industries.* Knoxville, Tenn.: Tennessee Valley Authority, 1977.

Singleton, Arthur L., Jr. *Sources of Nuclear Fuel.* Oak Ridge, Tenn.: USAEC, June 1968.

Taylor, Raymond W., and Samuel W. Taylor. *Uranium Fever.* New York: Macmillan, 1970.

Taylor, Vince. *How the U.S. Government Created the Uranium Crisis (and the Coming Uranium Bust).* Los Angeles, Calif.: Pan Heuristics, June 1977.

――――. *The Myth of Uranium Scarcity.* Los Angeles, Calif.: Pan Heuristics, April 1977.

Tippit, John H. "Federal Incentives to Uranium Mining." *Rocky Mountain Law Review* 27 (1955):457–468.

U.S. Atomic Energy Commission. *Nuclear Fuel Supply,* WASH-1242. Washington: USAEC, May 1973.

――――. *Nuclear Industry Fuel Supply Survey,* WASH-1196. Washington: USAEC, 1972.

――――. *Survey of United States Uranium Marketing Activity,* WASH-1196(73). Washington: USAEC, 1973.

――――. *Survey of United States Uranium Marketing Activity,* WASH-1196(74). Washington: USAEC, 1974.

――――. *Uranium Exploration Expenditures in 1971 and Plans for 1972–73,* GJO-103(72). Washington: USAEC, 1972. Also subsequent years: 1973, 1974.

U.S. Congress. Joint Committee on Atomic Energy. *Private Ownership of Special Nuclear Materials.* Hearings, 88th Cong., 1st Sess., July 30, 31, and August 1, 1963. Washington: GPO, 1963.

――――. *Private Ownership of Special Nuclear Material.* Hearings, 88th Cong., 2d Sess., June 9, 10, 11, 15, and 25, 1964. Washington: GPO, 1964.

――――. *Problems of the Uranium Mining and Milling Industry.* Hearings,

85th Cong., 2d Sess., February 19, 24, and 25, 1958. Washington: GPO, 1958.

―――. Subcommittee on Raw Materials. *AEC Raw Material Program.* Hearings, 87th Cong., 1st Sess., November 15–16, 1961. Washington: GPO, 1961.

―――. *AEC Uranium Procurement Program.* Hearings, 87th Cong., 2d Sess., June 18–19, 1962. Washington; GPO, 1962.

―――. Report, June 1952. Washington: GPO, 1952.

U.S. Energy Research and Development Administration. Grand Junction Office. *Statistical Data of the Uranium Industry,* GJO-100(75). Grand Junction, Colo.: USERDA, 1976. Also subsequent years: 1976, 1977.

―――. Division of Production and Materials Management. *Survey of United States Uranium Marketing Activity, April 1975,* ERDA-24. Washington: USERDA, 1975.

―――. Division of Nuclear Fuel Cycle and Production. *Survey of United States Uranium Marketing Activity, April 1976,* ERDA 76-46. Springfield, Va.: NTIS, 1976.

―――. Division of Uranium Resources & Enrichment. *Survey of United States Uranium Marketing Activity, May 1977,* ERDA 77-46. Springfield, Va.: NTIS, 1977.

U.S. Energy Research and Development Administration. "Testimony before the New York Public Service Commission in Case 26974 (nuclear and fossil economics)." Washington: USERDA, 1977.

U.S. General Accounting Office. *Review of Selected Aspects of the Domestic Uranium Procurement Program, Atomic Energy Commission, June 1961.* Washington: USGAO, March 1962.

Woodmansee, Walter C. "Uranium." In U.S. Department of the Interior, Bureau of Mines, *Minerals Yearbook, 1970.* Washington: GPO, 1972. Also subsequent years: 1971, 1972, 1973, 1974.

―――. "Uranium." In U.S. Department of the Interior, Bureau of Mines, *Mineral Facts and Problems.* Washington: GPO, 1976.

The World Market

Australian Atomic Energy Commission. *Twenty-third Annual Report.* Being the Commission's Report for the Year Ended June 30, 1975.

―――. *Uranium in Australia.* A Collection of Articles on the History and Development of the Uranium Mining Industry in Australia. Sydney: AAEC, 1962.

Bergsten, C. Fred. "The Threat Is Real." *Foreign Policy* 14 (1974): 84–90.

Carlsson, Owe. "The Ranstad Project and Other Swedish Projects and

Possibilities." In *Uranium Supply and Demand*, ed. by M.J. Spriggs and K.D. Casteel. London: Mining Journal Books, 1977.

Chayes, Abram. "U.S. Non-proliferation Policies." In *Uranium Supply and Demand*, ed. by M.J. Spriggs and K.D. Casteel. London: Mining Journal Books, 1977.

Couture, Jacques. "The Effect of Reprocessing on Long-term Uranium Requirements." In *Uranium Supply and Demand*, ed. by M.J. Spriggs and K.D. Casteel. London: Mining Journal Books, 1977.

Dahlkamp, Franz J. "Uranium Exploration Techniques—Their Applicability and Limitations." In *Uranium Supply and Demand*, ed. by M.J. Spriggs and K.D. Casteel. London: Mining Journal Books, 1977.

Darnley, A.G., E.M. Cameron, and K.A. Richardson. "The Federal-Provincial Uranium Reconnaissance Program." In *Uranium Exploration '75*, by Geological Survey of Canada. Ottawa: Supply and Services Canada, 1976.

Donndorf, H. Michael. "Co-operative Ventures with Developing Countries—An Assessment of Recent Experience." In *Uranium Supply and Demand*, ed. by M.J. Spriggs and K.D. Casteel. London: Mining Journal Books, 1977.

Duval, Bernard C. "The Changing Picture of Uranium Exploration." In *Uranium Supply and Demand*, ed. by M.J. Spriggs and K.D. Casteel. London: Mining Journal Books, 1977.

Gall, Norman. "Atoms for Brazil, Dangers for All." *Foreign Policy* 23 (1976):155–201.

Grey, Anthony J. "Australian Uranium—Will It Ever Become Available?" In *Uranium Supply and Demand*, ed. by M.J. Spriggs. London: Uranium Institute, 1976.

———. "Current Australian Uranium Position." In *Uranium Supply and Demand*, ed. by M.J. Spriggs and K.D. Casteel. London: Mining Journal Books, 1977.

Griffith, J.W. *The Uranium Industry—Its History, Technology and Prospects*. Ottawa: Queen's Printer, 1967. Mineral Report 12 of the Mineral Resources Division, Department of Energy, Mines and Resources, Ottawa.

Harris, Stuart. Statement of Evidence to the Ranger Uranium Environmental Enquiry, Canberra, June 29, 1976.

———. *Some Aspects of the Economics of the Uranium Industry in Australia*, CRES Report R/R1. Canberra: Australian National University, Centre for Resource and Environmental Studies, 1976, pp. 212–224.

Haynal, George. "Canadian Safeguards Policies." In *Uranium Supply and Demand*, ed. by M.J. Spriggs and K.D. Casteel. London: Mining Journal Books, 1977.

Kostuik, John. "The Uranium Industry." *Energy Policy* (September 1976):212–224.

Krasner, Stephen D. "Oil Is the Exception." *Foreign Policy* 14 (1974): 68–83.

MacNabb, G.M. "North American Uranium Resources: Policies, Prospects and Pricing." In *Uranium Supply and Demand,* ed. by M.J. Spriggs. London: Uranium Institute, 1976.

Mandel, H. "Uranium Demand and Security of Supply—A Consumer's Point of View." In *Uranium Supply and Demand,* ed. by M.J. Spriggs. London: Uranium Institute, 1976.

McKnight, Allan. "World Uranium Supplies." *SIPRI Yearbook 1972: World Armaments and Disarmament.* Stockholm: Almqvist and Wiksell, 1972, pp. 470–482.

Merlin, Harold B. "Canada's Uranium Production Potential." In *Uranium Supply and Demand,* ed. by M.J. Spriggs and K.D. Casteel. London: Mining Journal Books, 1977.

Messer, Klaus-Peter, "Uranium Demand as Judged by Electric Utilities." In *Uranium Supply and Demand,* ed. by M.J. Spriggs and K.D. Casteel. London: Mining Journal Books, 1977.

Mikdashi, Zuhayr. "Collusion Could Work." *Foreign Policy* 14 (1974):57–67.

Mizumachi, Kyoji. "Nuclear Anxieties and Their Impact on Nuclear Power Programmes." In *Uranium Supply and Demand,* ed. by M.J. Spriggs and K.D. Casteel. London: Mining Journal Books, 1977.

OECD Nuclear Energy Agency and International Atomic Energy Agency. *Uranium: Resources, Production, and Demand.* December 1975. Paris: OECD, 1975.

Patterson, J.A. "Foreign Resources and Production Capability." In *Uranium Industry Seminar,* GJO-108(75). Grand Junction, Colo.: USERDA, 1975.

Petit, A. "Economic and Political Environment of the Uranium Mining Industry." In *Uranium Supply and Demand,* ed. by M.J. Spriggs. London: Uranium Institute, 1976.

Potter Partners. *The Australian Uranium Industry.* Melbourne: Potter Partners, 1977.

Ruzicka, V. "New Sources of Uranium? Types of Uranium Deposits Presently Unknown in Canada." In *Uranium Exploration '75,* by Geological Survey of Canada. Ottawa: Supply and Services Canada, 1976.

Stephany, Manfred. "The Impact of Nonproliferation Policies on the Uranium Market—An Industry View." In *Uranium Supply and Demand,* ed. by M.J. Spriggs and K.D. Casteel. London: Mining Journal Books, 1977.

Stockholm International Peace Research Institute. *World Armaments and Disarmament: SIPRI Yearbook 1977.* Cambridge, Mass.: M.I.T. Press, 1977.

Tilley, C.E. "Raw Materials for Atomic Power." In *Atomic Energy,* ed. by J.L. Crammer and R.E. Peierls. Harmondsworth, England: Pelican, 1950.

U.S. Congress. House. Committee on Interstate and Foreign Commerce. Subcommittee on Oversight and Investigations. *International Uranium Cartel.* Hearings, 95th Cong., 1st Sess., May 2, June 10, 16, 17, and August 15, 1977. Washington: GPO, 1977.

U.S. Congress. Jont Committee on Atomic Energy. *Proposed Modification of Restrictions on Enrichment of Foreign Uranium for Domestic Use.* Hearings, 93d Cong., 2d Sess., September 17–18, 1974. Washington: GPO, 1975.

————. Raw Materials Subcommittee. *Report on a Visit to Australia.* February 9, 1955. Washington: GPO, 1955.

U.S. General Accounting Office. *Certain Actions That Can Be Taken To Improve This Nation's Uranium Picture,* EMD-76-1. Washington: USGAO, July 2, 1976.

von Kielin, A. "Alternative Sources of Supply." In *Uranium Supply and Demand,* ed. by M.J. Spriggs. London: Uranium Institute, 1976.

Williams, R.M. "Uranium to 2000, An Exploration Challenge." In *Uranium Exploration '75,* by Geological Survey of Canada. Ottawa: Supply and Services Canada, 1976.

————. *Uranium Supply to 2000: Canada and the World.* Mineral Bulletin MR 168. Ottawa: Supply and Services Canada, 1976.

———— and H.W. Little. *Canadian Uranium Resource and Production Capability.* Mineral Bulletin MR 140. Ottawa: Supply and Services Canada, 1973.

Williams, R.M., H.W. Little, W.A. Gow, and R.M. Berry. *L'uranium et le thorium au Canada: ressources, production et potentiel.* Mineral Bulletin MR 117F. Ottawa: Supply and Services Canada, 1972.

Willrich, Mason, and Philip M. Marston. "Prospects for a Uranium Cartel." *Orbis* XIX,1 (Spring 1975):166–185.

Wonder, Edward F. "Nuclear Commerce and Nuclear Proliferation: Germany and Brazil, 1975." *Orbis* 21,1 (Summer 1977):277–306.

————. *Nuclear Fuel and American Foreign Policy: Multilateralization for Uranium Enrichment.* Boulder, Colo.: Westview Press, 1977.

Worroll, R.E. "The Pattern of Uranium Production in South Africa." In *Uranium Supply and Demand,* ed. by M.J. Spriggs. London: Uranium Institute, 1976.

Wright, Robert J. "Foreign Uranium Developments." In USERDA, *Uranium Industry Seminar,* GJO-108(76). Grand Junction, Colo.: USERDA, 1976.

Index

Index

About the Author

Charles F. Zimmermann received the bachelor's degree from Brown University in 1971 and the Ph.D. in Agricultural Economics from Cornell University in 1978. He is presently a consultant at Foster Associates, Inc. in Washington, D.C. In addition to his professional work on the economics of energy supply and antitrust matters, another area of his research interest is nuclear power safety and environmental issues. He has coauthored an article with Robert Pohl on "The Potential Contribution of Nuclear Energy to U.S. Energy Requirements," appearing in *Energy* in 1977.